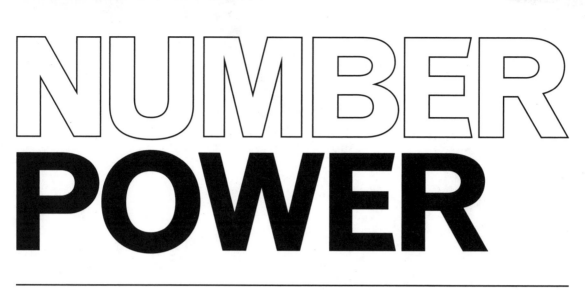

NUMBER POWER

A REAL WORLD APPROACH TO MATH

Word Problems

KENNETH TAMARKIN

McGraw Hill **Education**

Bothell, WA • Chicago, IL • Columbus, OH • New York, NY

www.mheonline.com

 **Education**

Send all inquiries to:
Contemporary/McGraw-Hill
130 E. Randolph, Suite 400
Chicago, IL 60601

ISBN: 978-0-07-659231-9
MHID: 0-07-659231-6

Printed in the United States of America.

2 3 4 5 6 7 8 9 10 RHR 15 14 13 12 11

TABLE OF CONTENTS

Multiplication and Division Word Problems: Whole Numbers

Multiplication and Division Word Problems: Decimals and Fractions

Using Proportions

Strategies with Mixed Word Problems

Percent Word Problems

Combination Word Problems

Posttest A 156

USING NUMBER POWER 161

Posttest B 177

TO THE STUDENT

Welcome to *Word Problems:*

This Number Power workbook is designed to help you understand how to solve word problems found in the workplace and in your everyday experiences. The first section of the book, Building Number Power, provides step-by-step instruction and practice in reading and solving word problems. The last section of the book, Using Number Power, provides further practice in solving word problems in real-life situations.

Throughout this book, you will be encouraged to use mental math. You will see word problems with small or easy to compute numbers. Once you figure out which math operation you should use, you might be able to do the math in your head.

Learning how and when to use a calculator is also an important skill to develop. In real life, problem solving often involves the smart use of a calculator—especially when working with large numbers. You need to know when an answer on a calculator makes sense, so be sure to have an estimated answer in mind. Remember that a calculator can help you only if you set up the problem and use the calculator correctly.

To get the most from your work, do each problem carefully. Check each answer to make sure you are working accurately. An answer key is provided at the back of the book. Inside the back cover is a chart to help you keep track of your score on each exercise.

Also included in the back of the book are quick reference pages for formulas and measurements. Refer to these pages for a quick review and resource.

Pretest

This test will tell you which sections of *Number Power Word Problems* you need to concentrate on. Do every problem that you can. Round decimals to the nearest cent or the nearest hundredth. Correct answers are listed by page number at the back of the book. After you check your answers, the chart at the end of the test will guide you to the pages of the book where you need additional work.

1. 1,600 pounds of steel are used to make a subcompact car. The automobile plant produced 840 identical subcompact cars in one day last week. How many pounds of steel were needed that day to make the subcompact cars?

   a. 2,440 pounds

   b. 1,344,000 pounds

   c. 760 pounds

   d. 244,000 pounds

   e. none of the above

2. At the end of one year, Melinda had grown $2\frac{1}{4}$ inches to a height of $48\frac{3}{8}$ inches. What was her height at the beginning of the year?

   a. $50\frac{5}{8}$ inches

   b. $46\frac{1}{8}$ inches

   c. $43\frac{1}{2}$ inches

   d. $21\frac{1}{2}$ inches

   e. $108\frac{27}{32}$ inches

3. Cynthia took 19 friends roller skating. If it cost $7.50 for each friend to get in and $5.00 for each of them to rent inline skates, how much money did Cynthia have to collect?

   a. $31.50

   b. $12.50

   c. $95.00

   d. $142.50

   e. $237.50

4. During the big spring sale, Oksana bought a coat for $79.50, which was 75% of the original price. What was the original price of the coat?

   a. $106.00

   b. $59.63

   c. $154.50

   d. $94.34

   e. not enough information given

5. Diana makes lemonade from the powdered concentrate by combining 5 tablespoons of concentrate with 2 cups of water. The directions say you should use 24 cups of water for the entire container of concentrate. How many tablespoons of concentrate are in the container?

   a. 240 tablespoons

   b. 130 tablespoons

   c. 110 tablespoons

   d. 60 tablespoons

   e. 31 tablespoons

6. Mei was told that she would have to pay $684 interest on a $3,600 loan. What interest rate would she have to pay?

   a. $2,912

   b. $4,284

   c. 19%

   d. 5.3%

   e. 81%

7. Jack bought a turkey for $10.34 and a chicken for $5.17. How much did he spend?

   a. $2.00

   b. $5.17

   c. $15.51

   d. $53.46

   e. $20.00

8. The New Software Company received a shipment of 200,000 foam pellets to be used in packing boxes. If New Software uses on the average 400 pellets for each box, how many boxes can be packed using the shipment of pellets?

   a. 2,000 boxes

   b. 800 boxes

   c. 199,600 boxes

   d. 80,000,000 boxes

   e. 500 boxes

9. Oranges cost $2.40 a dozen. Wynona bought the fruit pictured here. How much money did she spend on the oranges?

   a. $7.20

   b. $2.44

   c. $2.36

   d. $0.80

   e. $0.60

10. A roast weighing 3.15 pounds is cut into 24 slices. On the average, how much does each slice weigh?

   a. 27.15 pounds

   b. 20.85 pounds

   c. 75.60 pounds

   d. 0.13 pound

   e. 1.31 pounds

11. Barbara needed 180 inches of masking tape to mask a window for painting. How many rolls of masking tape does she need to mask 12 identical windows?

   a. 18 feet

   b. 15 rolls

   c. 192 inches

   d. 168 inches

   e. not enough information given

12. A factory produces $\frac{7}{8}$-ton steel girders. How much steel does the factory need to produce 600 of the girders?

   a. 525 tons

   b. 52.5 tons

   c. 686 tons

   d. 68.6 tons

   e. none of the above

13. During a sale, Nikki bought a sports jacket that was reduced by $98 to $190. What was the original price of the sports jacket?

   a. $92

   b. $288

   c. $276

   d. $96

   e. $291

14. Out of 1,400 people polled, 68% were in favor of a nuclear arms freeze, 25% were against it, and the rest were undecided. How many people were undecided?

   a. 93 people

   b. 350 people

   c. 952 people

   d. 98 people

   e. 1,307 people

15. A $1\frac{1}{4}$-pound lobster costs $7.80. How much does it cost per pound?

    a. $9.75

    b. $6.24

    c. $6.55

    d. $9.05

    e. $1.56

16. At the gasoline station, Erica tried to fill up her 12-gallon gasoline tank. When the tank was full, the gasoline pump looked like the picture at the right. How much gasoline was in the tank before Erica started pumping?

    a. 8.71 gallons

    b. 10.07 gallons

    c. 5.22 gallons

    d. 10.30 gallons

    e. 18.78 gallons

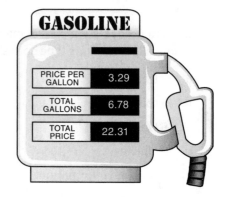

17. During the Washington's Birthday Clearance Sale, Gayle bought a $96 coat that was reduced by $\frac{1}{3}$. What was the sale price of the coat?

    a. $32

    b. $64

    c. $288

    d. $93

    e. none of the above

18. Sherry's rent has been increased $65 a month to $780 a month. What had she been paying?

    a. $715

    b. $845

    c. $72

    d. $9,360

    e. $50,700

19. Lily's allergy pills come in the bottle pictured at the right. She takes four tablets a day. How many tablets did she have left after taking the tablets for 30 days?

    a. 130 tablets

    b. 216 tablets

    c. 120 tablets

    d. 370 tablets

    e. not enough information given

20. A cereal manufacturer puts 2 ounces of sugar in every box of cereal. How many pounds of sugar are needed for 1,000 boxes?

    a. 50 pounds

    b. 20 pounds

    c. 125 pounds

    d. 200 pounds

    e. 625 pounds

21. For a survey to be considered valid, 15% of the 6,000 questionnaires had to be returned. At least how many questionnaires had to be returned?

    a. 900 questionnaires

    b. 400 questionnaires

    c. 40,000 questionnaires

    d. 5,985 questionnaires

    e. not enough information given

22. An oil truck carried 9,008 gallons of oil. After making seven deliveries averaging 364 gallons each, how much oil was left in the truck?

    a. 174 gallons

    b. 9,379 gallons

    c. 8,644 gallons

    d. 6,460 gallons

    e. 8,637 gallons

23. The Custom Tailor Shop used $2\frac{5}{8}$ yards of fabric to make a size 6 dress and $2\frac{3}{4}$ yards of the same fabric to make the same dress in size 8. How much fabric was needed to make both dresses?

    a. $\frac{1}{8}$ yard

    b. $\frac{5}{8}$ yard

    c. $5\frac{3}{8}$ yards

    d. $7\frac{1}{4}$ yards

    e. 14 yards

24. How much would a 1.62-pound package of top round steak cost at $2.43 a pound?

    a. $1.50

    b. $4.05

    c. $0.81

    d. $15.00

    e. $3.94

25. Money available for financial aid at Santa Carla Community College has dropped $462,000 from last year's $1,126,200. The college decided to divide the aid equally among 820 students who needed the money. How much did each student get in financial aid?

    a. $563.41

    b. $810.00

    c. $1,373.41

    d. $1,936.82

    e. none of the above

Questions 26–28 are based on the information below and to the right.

The Bargain Basement marks down clothes depending on how long they have been displayed. The day an item is put on the racks or in a bin, a date ticket and a price tag are attached to it. The chart shows the amount of discount if the date ticket is at least the listed number of days old.

Days	Markdown
10 days	10%
20 days	25%
30 days	40%
40 days	75%

26. On May 30, Sharon went shopping at the Bargain Basement. She found one top she wanted dated April 14 with a price tag of $18. How much did she have to pay for the top?

    a. 46 days

    b. $13.40

    c. $4.50

    d. $22.50

    e. $18.00

27. On July 20 at the Bargain Basement, Belquis selected five bathing suits. The red bathing suit was dated June 28 and had a $40 price tag. The floral print suit was dated July 8 and had a price tag of $30. The violet bathing suit was dated June 18 and had a price of $45. The striped bathing suit was dated July 15 and had a price of $28. The black bathing suit was dated June 2 and had a price of $60. Which bathing suit was the least expensive?

    a. the red bathing suit

    b. the floral print bathing suit

    c. the violet bathing suit

    d. the striped bathing suit

    e. the black bathing suit

28. On December 3 at the Bargain Basement in New Hampshire, Marcia bought a sweater dated November 28. She paid with a $50 bill. Since New Hampshire does not have a sales tax, she paid no tax. How much change did Marcia receive?

    a. $11.00

    b. $89.00

    c. $19.50

    d. $14.90

    e. not enough information given

29. Last month, Francisco used 445 kWh of electricity in his home. On his bill, there was a customer charge of $5.81, a delivery service charge of $0.047 per kWh, and a supplier service charge of $0.032 per kWh. Which expression could be used to calculate his total bill?

    a. 445 kWh ($5.81 + $0.047 + $0.032)

    b. $5.81 + $0.047 + $0.032

    c. $5.81 + 445 kWh ($0.047 + $0.032)

    d. 445 kWh + $5.81 ($0.047 + $0.032)

    e. 445 kWh ($0.047 + $0.032)

30. MovieFlix offers a flat rate of $8.99 a month for unlimited movie rentals. ApCo rents movies for $2.99 each. What is the greatest number of movies Noah and Juno could watch in a month for which ApCo offers the better deal?

    a. 1 movie

    b. 2 movies

    c. 3 movies

    d. 4 movies

    e. 6 movies

PRETEST CHART

If you miss more than one problem in any section of this test, you should complete the lessons on the practice pages indicated on this chart. If you do not miss any problems in a section of this test, you may not need further study in this chapter. However, to master solving word problems, we recommend that you work through the entire book. As you do, focus on the skills covered in each chapter.

Problem Numbers	Skill Area	Practice Pages
13, 18	add or subtract whole numbers	17–36
1, 8	multiply or divide whole numbers	57–71
2, 23	add or subtract fractions	48–56
12, 15	multiply or divide fractions	75–83
7, 16	add or subtract decimals	37–47, 52–56
10, 24	multiply or divide decimals	72–74, 82–83
4, 6, 21	percents	112–129
9, 20	conversion	94–95
11, 28	not enough information given	104–108
3, 5, 14, 17, 19, 22, 25, 26, 27, 29, 30	multistep word problems	130–155

BUILDING
NUMBER
POWER

INTRODUCTION TO WORD PROBLEMS

Steps in Solving Word Problems

A **word problem** is a sentence or group of sentences that tells a story, contains numbers, and asks the reader to answer a question based on the numbers.

This is an example of a word problem:

> Last week Paula earned $694. The week before, she earned $588. What was the total amount of money she earned?

In this book, you will use five steps to solve word problems. It is important to follow these steps to organize your thinking. They will help you figure out what may seem to be a difficult puzzle. In each case, read the problem carefully, more than once if necessary. Then follow these steps.

STEP 1   Decide what the *question* is asking you to find.

STEP 2   Then decide what *information* is *necessary* in order to solve the problem.

STEP 3   Next, decide what *arithmetic operation* to use.

STEP 4   Work out the problem and find the solution. Check your arithmetic.

STEP 5   Finally, *reread the question* to make sure that your answer *is sensible*.

Many people can do some word problems in their heads. This is known as **math intuition** and works well with small whole numbers. Math intuition often breaks down with larger numbers, decimals, and especially fractions. Additionally, word problems of two or more steps can be even more difficult.

You should practice the five-step approach even with problems that you could solve in your head. Then you will have something to fall back on when intuition is not enough.

Step 1: The Question

After reading a word problem, the first step in solving it is to decide what is being asked for. You must find the question.

The following word problem consists of only one sentence. This sentence asks a question and contains the information needed to solve the problem.

EXAMPLE 1   How much did Maxim spend on dinner when the food cost $20 and the tax was $1?

The question asks, "How much did Maxim spend on dinner?"

The next word problem contains two sentences. One sentence asks the question, and the other sentence gives the information necessary to solve the problem.

EXAMPLE 2   Mary got $167 a month in food stamps for 9 months. What was the total value of the stamps?

The question asks, "What was the total value of the stamps?"

Example 3 also contains two sentences. Notice that *both* sentences contain information necessary to solve the problem.

EXAMPLE 3   The Little Sweetheart tea set normally costs $18.95. How much does Alicia save by buying the tea set for her daughter on sale for $10.49?

The question asks, "How much does Alicia save?"

Sometimes the question does *not* have a question mark.

EXAMPLE 4   Fredda has $27 in her checking account. She writes checks for $15 and $20. Find how much money she needs to deposit in order to cover the checks.

The question asks, "Find how much money she needs to deposit in order to cover the checks."

**Underline the question in each of the following word problems.
DO NOT SOLVE!**

1. Last winter it snowed 5 inches in December, 17 inches in January, 13 inches in February, and 2 inches in March. How much snow fell during the entire winter?

2. To cook the chicken, first brown it for 10 minutes. Then lower the temperature and let it simmer for 20 more minutes. What is the total cooking time?

3. Find the cost of parking at the meter for 3 hours if it costs 25 cents an hour to park.

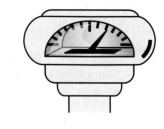

4. How many years did Joe serve in prison, if his sentence of five years was reduced by three for good behavior?

**In each of the following word problems, the question is missing.
Write a possible question for each word problem.**

5. Elina loves to garden. She has $30 to buy flats of annuals. Each flat costs $4.98.

6. The recycling plant paid $22 a ton for recycled newspaper. The city of Eugene delivered 174 tons of newspaper to the recycling plant.

7. A factory needs $2\frac{1}{3}$ yards of material to make a coat. The material comes in 60-yard rolls.

8. A tablet of Extra-Strength Tylenol contains 500 milligrams of acetaminophen. The normal adult dose is 2 tablets every 4 to 6 hours, not to exceed 8 tablets in 24 hours.

9. Computer Connection slashed the price of its best selling single lens reflex digital camera from $799 to $549.

10. The Arctictec textile factory, which operates round the clock, can produce 42,000 yards of Arctictec fabric in 24 hours.

11. In a normal week, Great Deal Used Cars sells 38 cars. During the big Presidents' Week sale, Great Deal sold 114 cars.

12. The snack bar menu, shown at the right, was posted on the wall.

MENU	
Super Pretzels	$1.50
Chips	$0.65
Nachos	$2.25
Soda (12 fl oz can)	$1.20
Hot Dog	$2.50
Hamburger	$3.50
Pizza Slice	$2.75
Milk	$1.50

Step 2: Selecting the Necessary Information

After finding the question, the next step in solving a word problem is selecting the **necessary information.** The necessary information consists of the **numbers** and the **labels** (words or symbols) that go with the numbers. The necessary information includes *only* the numbers and labels that you need to solve the problem.

The labels make the numbers in word problems concrete. For example, the necessary information in Example 1 below includes not just the number *5*, but includes *5 shirts*. Paying close attention to labels will help you learn many of the methods shown in this book and will help you avoid common mistakes with word problems.

After each of the following examples, the necessary information is listed.

EXAMPLE 1 A shirt costs $9.99. What is the cost of 5 shirts?

The necessary information is *$9.99* and *5 shirts*. The labels are the dollar sign (*$*) and the word *shirts*.

EXAMPLE 2 Four fluid ounces of detergent are needed to clean a load of laundry. How many more loads of laundry can you clean if you buy the large bottle of detergent rather than the small bottle shown at the right?

The necessary information is *4 fluid ounces*, *64 fluid ounces*, and *96 fluid ounces*. The numbers are followed by the label words *fluid ounces*. Note that you get some of the necessary information from the pictures.

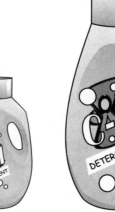

64 fluid ounces   96 fluid ounces

In each word problem, find the necessary information. Circle the numbers and underline the labels. Then write the label that would be a part of the answer, but DO NOT SOLVE!

1. On Friday a commuter train took 124 commuters to work and 119 commuters home. How many commuters rode the train that day?

2. There are 14 potatoes in the bag at the right. What is the average weight of each potato?

3. Frank bought the package of loose-leaf paper shown below and put 60 pages in his binder. How many pages were left in the package?

Potatoes
5 lb

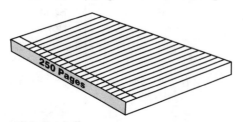

250 Pages

Necessary vs. Given Information

Sometimes a word problem contains numbers that aren't needed to answer the question. You must read problems carefully to choose only the necessary information.

Notice this important difference: The **given information** includes *all* the numbers and labels in a word problem.

The **necessary information** includes *only* those numbers and labels needed to solve the problem.

EXAMPLE 1   Nelson travels to and from work with 3 friends every day. The round trip is 9 miles. If he works 5 days a week, how many miles does he commute in a week?

given information: 3 friends, 9 miles, 5 days
necessary information: 9 miles, 5 days

To figure out how many miles he commutes in a week, you do not need to know that Nelson travels with 3 friends.

EXAMPLE 2   There are 7,000 people living in Dry Gulch. Of the 3,000 people who are registered to vote, only 1,700 people participated in the last election. How many registered voters did not vote?

given information: 7,000 people, 3,000 people, 1,700 people
necessary information: 3,000 people, 1,700 people

All of the numbers have the same label—*people*. However, the total number of people in the town (7,000) is not needed.

Sometimes you will have to choose necessary information from a chart or picture containing other information as well.

EXAMPLE 3   According to the chart, how many hours did Eduardo work on Friday and Saturday?

Hours Worked	
Monday	4
Wednesday	2
Friday	6
Saturday	8

given information: 4 hours, 2 hours, 6 hours, 8 hours
necessary information: 6 hours, 8 hours

In this book, you will practice choosing information from charts and pictures.

This exercise will help you tell the difference between *given* and *necessary* information. Underline the given information. Circle the necessary information. DO NOT SOLVE!

1. Mona is 22 years old. She has a sister who is 20 years old and a boyfriend who is 23. How much older is Mona than her sister?

2. Rena receives $386 a month from TANF. She also receives $167 a month in food stamps in order to help feed her two children. How much public assistance does she receive each month?

3. Marilyn works three times as many hours as her 20-year old sister Laura. Laura works 10 hours a week. How many hours a week does Marilyn work?

4. Suzanne has a 7-year old car. According to the chart at the right, how much does she spend on gasoline during the first 2 months of the year?

Gasoline Expenses	
January	$143
February	$139
March	$140
April	$131

5. During the winter, the Right Foot shoe store spent $2,400 for oil heat and sold $135,800 worth of shoes. If oil costs $3.60 per gallon, how many gallons did the shoe store buy?

6. Gia, who is 45 years old, cooks dinner for the eight people in her family. Her husband, Jack, cooks breakfast in the morning for only half of the family. For how many people does Jack cook?

7. In a factory of 4,700 workers, 3,900 are skilled laborers. Of the employees, 700 people are on layoff. How many people are currently working at the factory?

8. Maritza buys the bottle of cola shown at the right for $2.49. How many 12-ounce glasses can she fill from the bottle?

Cola

64 fl oz

Finding Addition Key Words

In the first chapter, you worked on finding the question and the necessary information in a word problem. The third step in solving a word problem is **deciding which arithmetic operation to use.**

You will now look at word problems that can be solved by using either addition or subtraction. In this book, you will learn five methods to decide whether to add or subtract.

> **1.** finding the key words
> **2.** restating the problem
> **3.** making drawings and diagrams
> **4.** writing number sentences
> **5.** using algebra

You will also work with making estimates and substitutions.

All of these methods are useful in understanding and solving word problems. After learning them, you may decide to use one or more of the methods that you find most helpful.

How do you know that you must add to solve a word problem? **Key words** can be helpful. A key word is a clue that can help you decide which arithmetic operation to use.

> **Note:** *How many, how much,* and *what* are general mathematics question words, but they are not key words. They help to identify the question but do not tell you whether to add, subtract, multiply, or divide.

The following examples contain addition key words.

EXAMPLE 1 **What is the sum of $3 and $2?**

addition key words: sum, and

The sum is the answer to an addition problem. Therefore, when the word *sum* appears in a word problem, it is a clue that you should probably add to solve the problem.

EXAMPLE 2 **The small cup contains 16 fluid ounces of soda. The large cup contains 6 more fluid ounces. How many fluid ounces are in the large cup?**

addition key word: more

The word *more* suggests that you should add the two amounts together.

**In the following exercise, circle the key word(s) that suggest addition.
DO NOT SOLVE!**

1. Karen bought a new car for $15,640 plus $4,600 for options. How much did she spend for the car?

2. Julia bought four lemons and twelve oranges. How many pieces of fruit did she buy altogether?

3. Although she has lost 15 pounds, Marika wants to lose 10 more pounds. How much weight does she want to lose altogether?

4. The price of a $6 general admission ticket to the ballpark will increase by $1 next year. What will be the general admission price next year?

5. A recipe for pumpkin pie says that an extra 2 tablespoons of sugar can be added for extra sweetness. The standard recipe is below. How many tablespoons of sugar are needed for a sweeter pie?

   GRANNY'S PUMPKIN PIE
   1 tsp. salt
   4 tablespoons sugar
   2 tsp. cinnamon

6. When Nadia did laundry, she found the coins pictured below. How much change did she find in all?

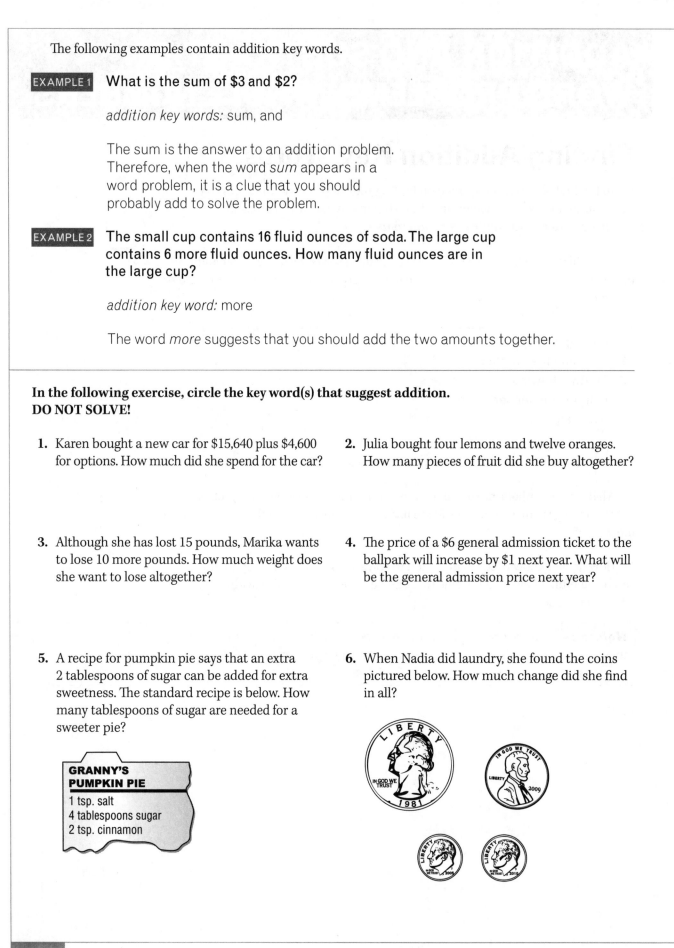

Solving Addition Word Problems with Key Words

You have now looked at the first three steps in solving a word problem.

STEP 1	Finding the question
STEP 2	Selecting the necessary information
STEP 3	Deciding what arithmetic operation to use

The next step in solving a word problem is doing the arithmetic. People who have not learned to think carefully about word problems may rush into doing the arithmetic and become confused. However, the three steps before doing the arithmetic and the one step after provide a good way to organize your thinking to solve a problem. The actual arithmetic is only one of several necessary steps.

Sometimes the arithmetic can be done in your head as **mental math.** Here are some examples that can be done as mental math.

EXAMPLE   The small cup contains 16 fluid ounces of soda. The large cup contains 6 more fluid ounces. How many fluid ounces are in the large cup?

STEP 1   *question:* How many fluid ounces are in the large cup?

STEP 2   *necessary information:* 16 fluid ounces, 6 fluid ounces

STEP 3   *addition key word:* more

STEP 4   *add:* 16 fluid ounces + 6 fluid ounces = **22 fluid ounces**

$$\begin{array}{r} 16 \\ +\ 6 \\ \hline 22 \end{array}$$

Once you have completed the arithmetic, there is one last step: reread the question and make sure that the answer is sensible.

For instance, if you had subtracted, you would have gotten an answer of 10 fluid ounces. Would it have made sense to say that the smaller cup contains 16 fluid ounces and the larger cup contains 10 fluid ounces?

For each problem, circle the key word(s) and do the arithmetic. Be sure to include the label as part of your answer.

1. After 5 inches of snow fell on a base of 23 inches of snow, how many inches of snow were on the ski trail altogether?

2. What is the total weight of a 4,500-pound truck carrying an 836-pound load?

3. According to the price chart at the right, how much does it cost to buy a sofa and a reclining chair?

Furniture	
Sofa	$529
Love seat	$319
Reclining chair	$449
Coffee table	$299

Finding Subtraction Key Words

Each of the key words in the previous exercises helped you decide to add. Other key words may help you decide to subtract. Here is an example of a word problem using subtraction key words.

EXAMPLE The large cup contains 16 fluid ounces of soda. The small cup contains 6 fluid ounces less than the larger cup. How many fluid ounces does the small cup contain?

      STEP 1 *question:* How many fluid ounces does the small cup contain?

      STEP 2 *necessary information:* 16 fluid ounces, 6 fluid ounces

      STEP 3 *subtraction key words:* less than

Note: In some subtraction word problems, the key words *less* and *than* are separated by other words. This is also true for *more* and *than*.

In the following exercise, circle the key word(s) that suggest subtraction. DO NOT SOLVE!

1. Bargain Airlines is $25 cheaper than First Class Air. First Class Air charges $200 for a flight from Kansas City to St. Louis. What does Bargain Airlines charge?

2. Out of 7,103 students, State College had 1,423 graduates last year. This year there were 1,251 graduates. What was the decrease in the number of graduates?

3. The smaller steak weighs 5 ounces less than the larger steak. How much does the smaller steak weigh?

 12 oz

$4.59 $2.99

4. The large engine has 258 horsepower. The economy engine has 92 horsepower. What is the difference in the horsepower between the two engines?

5. This year Great Rapids has 15 schools. Next year the number of schools will be reduced by 2. How many schools does the city plan to open next year?

Solving Subtraction Word Problems with Key Words

You can now complete the example in the "Finding Subtraction Key Words" lesson.

EXAMPLE The large cup contains 16 fluid ounces of soda. The small cup contains 6 fluid ounces less than the larger cup. How many fluid ounces does the small cup contain?

STEP 1 *question:* How many fluid ounces does the small cup contain?

STEP 2 *necessary information:* 16 fluid ounces, 6 fluid ounces

STEP 3 *subtraction key words:* less than

STEP 4 *subtract:* 16 fluid ounces − 6 fluid ounces = **10 fluid ounces**

$$\begin{array}{r} 16 \\ -\ 6 \\ \hline 10 \end{array}$$

In each problem below, circle the key word(s) and do the arithmetic. Be sure to include the label as part of your answer.

1. How much change did Kam receive when he paid for $36 worth of gas with a $50 bill?

2. What is the difference in price between the two pre-owned cars below?

$12,635 $7,849

3. Harold weighs 161 pounds, and his wife Nora weighs 104 pounds. How much more does Harold weigh than Nora?

4. Caroline's heating bill was $215 in March and $46 in April. By how much did her heating bill decrease in April?

5. Luiz bought a 21-pound sirloin strip. After the butcher trimmed the fat and cut the strip into steaks, the weight of the meat was 17 pounds. How much less did the meat weigh after the fat was trimmed?

Key Word Lists for Addition and Subtraction

Now you know that some key words may help you decide to add. Other key words may help you decide to subtract.

Here are some important key words to remember. You may want to add more words to these lists.

ADDITION KEY WORDS

sum	raise
plus	both
add	combined
and	in all
total	altogether
increase	additional
more	extra

SUBTRACTION KEY WORDS

less than	left
more than	remain
decrease	fell
difference	dropped
reduce	change
lost	

nearer
farther } other -er comparison words

Solving Addition and Subtraction Problems with **Key Words**

Now that you have seen how key words are used in addition and subtraction word problems, look at the following examples to see the difference between addition word problems with key words and subtraction word problems with key words.

EXAMPLE 1 It snowed 7 inches on Monday and 5 inches on Friday. What was the total amount of snow for the week?

STEP 1 *question:* What was the total amount of snow?

STEP 2 *necessary information:* 7 inches, 5 inches

STEP 3 *addition key words:* and, total

STEP 4 *add:* 7 inches + 5 inches = **12 inches**

$$\begin{array}{r} 7 \\ + 5 \\ \hline 12 \end{array}$$

EXAMPLE 2 The city usually runs its entire fleet of 237 buses during the morning rush hour. On Thursday morning, 46 buses and 13 subway cars were out of service. How many buses were left for the Thursday morning rush hour?

STEP 1 *question:* How many buses were left Thursday morning?

STEP 2 *necessary information:* 237 buses, 46 buses
(13 subway cars is not necessary information.)

STEP 3 *subtraction key word:* left

STEP 4 *subtract:* 237 buses − 46 buses = **191 buses**

$$\begin{array}{r} 237 \\ - 46 \\ \hline 191 \end{array}$$

In addition word problems, numbers are often being combined, and you are looking for the total. In subtraction word problems, numbers are being compared, and you are looking for the difference.

In this exercise, circle the key word(s). Decide whether to add or to subtract. Then solve the problem.

1. The Hard to Find Books Web site sold 86 books on Monday and 53 books on Tuesday. How many books were sold altogether?

2. After selling 15 rings on Wednesday, a jeweler sold 31 rings and 4 necklaces on Thursday. How many more rings did she sell on Thursday than on Wednesday?

3. At the town meeting, votes are recorded on the vote tally board as shown at the right. What was the total number of votes?

Vote Tally Board	
Yes	564
No	365

4. This year the average daily attendance at the Southwest Art Museum was 2,530 people. Last year, the average daily attendance was 3,421 people. By how many people did the average daily attendance decrease this year?

5. Mammoth Oil advertises that with its new brand of oil, a car can be driven 10,000 miles between oil changes. With Mammoth's old oil, a car's oil had to be changed every 3,000 miles. How much farther can you drive with Mammoth's new oil than with its old oil?

6. Last year, the Gonzales family paid $830 a month for rent. If rent was increased by $35 a month, how much rent is the family paying now?

7. In April the Discount Appliance Outlet paid $426 for electricity. In July their bill rose to $542. How much more did they pay in July than in April?

Key Words Can Be Misleading

So far, you have seen one approach to solving word problems.

STEP 1   Find the question.

STEP 2   Select the necessary information.

STEP 3   Use the key words to decide what arithmetic operation to use.

STEP 4   Do the arithmetic.

This approach can work in many situations.

But Be Careful with Step 3!

Sometimes the same key word that helped you decide to add in one word problem can appear in another problem that requires subtraction.

The next two examples use the *same* numbers and the *same* key words. In Example 1, you must add to find the answer; in Example 2 on page 26, you must subtract.

EXAMPLE 1   Judy bought 4 cans of pineapple and 16 cans of applesauce. What was the total number of cans that she bought?

STEP 1   *question:* What was the total number of cans?

STEP 2   *necessary information:* 4 cans, 16 cans

$$\begin{array}{r} 16 \\ + 4 \\ \hline 20 \end{array}$$

STEP 3   *key words:* and, total
Since you are looking for a total, you should add.

STEP 4   4 cans + 16 cans = **20 cans**

EXAMPLE 2   Judy bought a total of 16 cans of fruit. Four were cans of pineapple. The rest were applesauce. How many cans of applesauce did she buy?

STEP 1   *question:* How many cans of applesauce did she buy?

$$\begin{array}{r} 16 \\ -\ 4 \\ \hline 12 \end{array}$$

STEP 2   *necessary information:* 4 cans, 16 cans

STEP 3   *key word:* total
Since you have been given a total and are being asked to find a part of it, you must subtract.

STEP 4   16 cans − 4 cans = **12 cans**

In both examples, the word *total* was used. In Example 1, the question asked you to find the total. Therefore, you had to add. In Example 2, however, the total (cans of fruit) was part of the information given in the problem. The question asked you to find the number of cans of applesauce, a part of the total. To do this, you had to subtract the number of cans of pineapple from the total number of cans.

These two examples show that key words can be good clues, **but they are only a guide to understanding a word problem.** If you use the key words without understanding what you are reading, you may end up doing the wrong arithmetic.

This exercise will help you examine more carefully problems containing key words. In each of the following exercises, the key word is left out and the solution is given. Two choices are given for the missing word; circle the correct one.

1. At 5 P.M. the temperature was 87 degrees. Since 2 P.M., the temperature _____ by 9 degrees. What was the temperature at 2 P.M.?

    87 degrees + 9 degrees = **96 degrees**   (dropped, rose)

2. It was 78 degrees at 11 A.M. At 2 P.M. the temperature _____ to 87 degrees. By how many degrees did the temperature change?

    87 degrees − 78 degrees = **9 degrees**   (dropped, rose)

3. The 5% sales tax is going to _____ by 1%. What will be the new sales tax?

   5% + 1% = **6%**   (increase, decrease)

4. The 5% sales tax is going to _____ by 1%. What will be the new sales tax?

   5% − 1% = **4%**   (increase, decrease)

5. Starting next month, the Jones family will receive $14 _____ a week for food stamps. Currently they get $87 a week. How much will they receive each week next month?

   $87 − $14 = **$73**   (more, less)

6. The Johnson family's food stamp allotment was reduced. They now receive $14 a week _____, or $87 for stamps. What was their weekly allotment before the reduction?

   $87 + $14 = **$101**   (more, less)

7. Gloria used to keep her thermostat at 72 degrees. To save energy, she _____ it by 6 degrees. What was the new temperature in her apartment?

   72 degrees − 6 degrees = **66 degrees**   (raised, lowered)

8. Magda's mother came to visit for the weekend. To make sure that her mother was comfortable, she _____ the thermostat to 70 degrees. Usually, the thermostat was set at 66 degrees. By how much did Magda change the temperature?

   70 degrees − 66 degrees = **4 degrees**   (lowered, raised)

9. Gail normally ate 2,400 calories a day. While on a special diet, she ate 1,100 calories _____. How many calories a day did she eat on her diet?

   2,400 calories + 1,100 calories = **3,500 calories**   (more, less)

Restating the Problem

Have you ever tried to help someone else work out a word problem? Think about what you do. Often, you read the problem with the person, then discuss it or put it in your own words to help the person see what is happening. You can use this method—**restating the problem**—to help yourself solve a problem.

Restating the problem can be especially helpful when the word problem contains no key words. Look at the following example:

EXAMPLE Susan has already driven her car 2,700 miles since its last oil change. She still plans to drive 600 miles before changing the oil. How many miles does she plan to drive between oil changes?

STEP 1 *question:* How many miles does she plan to drive between oil changes?

STEP 2 *necessary information:* 2,700 miles, 600 miles

STEP 3 Decide what arithmetic operation to use. Restate the problem in your own words: "I know the number of miles Susan has already driven and the number of miles more that she plans to drive. I need to find the total number of miles between oil changes. I should add."

STEP 4 2,700 miles + 600 miles = **3,300 miles** between oil changes

$$\begin{array}{r} 2,700 \\ +600 \\ \hline 3,300 \end{array}$$

STEP 5 It makes sense that she will drive 3,300 miles between oil changes since you are looking for a number larger than the 2,700 miles that she has already driven.

Try this method with the next exercise. Read the problem, and then restate it to yourself. In future work with particularly confusing word problems, you should try this method of talking to yourself to understand the problems.

Each word problem is followed by two short explanations. One gives you a reason to add to find the answer. The other gives you a reason to subtract to find the answer. Circle the correct explanation.
DO NOT SOLVE!

1. Margi's weekly food budget has increased by $12 over last year's to $141 a week. How much was her weekly food budget last year?

   a. The budget has increased since last year. Therefore, you add the two numbers.

   b. Her food budget has increased over last year's. The new, larger budget is given. Therefore, you subtract to find last year's smaller amount.

2. In the runoff election for mayor, Fritz Neptune got 14,662 votes, and Julio Cortez got 17,139 votes. How many votes were cast in the election?

   a. To find the total number of votes cast, you should add the two numbers given.

   b. To find the number of votes cast, you should subtract to find the difference between the two numbers.

3. The difference between first-class (the most expensive fare) and coach airfare is $1,188. If coach costs $406, how much does first class cost?

   a. To find the cost of first class, you subtract to find the difference between the two fares.

   b. First class costs more than coach. Since you are looking for the larger fare, you add the smaller fare to the difference between the two fares.

4. It costs $66,840 to run and maintain the town's swimming pool. During the year, $59,176 is collected from user fees for the pool, and the town government pays the rest of the cost. How much money does the town government have to pay?

   a. To find the total cost, you add the cost of running and maintaining the pool to the amount collected in user fees.

   b. You are given the total cost of running the pool and the part of the cost covered by user fees. To find the cost to the town government, you subtract.

5. After reading a 320-page novel, Danika read a 205-page history book. How many pages did Danika read?

    a. Since you are looking for the total number of pages, you add.

    b. To find the difference between the number of pages in the two books, you subtract.

6. After making 24 bowls, Claire made 16 plates. How many pieces did she make?

    a. Since the number of pieces includes the number of bowls and plates, you add them together.

    b. Since you are looking for a difference, you subtract the number of plates from the number of bowls.

7. A factory has produced 48,624 microwave ovens so far this year. The company expects to produce 37,716 microwave ovens during the rest of the year. What is the projected production of ovens for the year?

    a. To find the projected production for the year, you subtract the number of microwave ovens to be produced from the number of ovens that have been produced so far.

    b. To find the projected production for the entire year, you add the number of ovens already produced to the number of ovens that are expected to be produced.

8. Ling has a 50,000-mile warranty on her car. As of today, she has driven 34,913 miles. How many miles are left on the warranty?

    a. To find the total number of miles left on the warranty, you add the number of miles Ling has driven to the number of miles that the warranty covers.

    b. Since Ling's car is still under warranty, you subtract the miles she has already driven from the mileage that the warranty covers.

Using Pictures and Diagrams to Solve Word Problems

Another approach that people use to solve word problems is to form a picture of the problem. While some people can do this in their heads, many people find it very useful to draw a picture or diagram of the problem.

EXAMPLE 1 Neely's Bar & Grill uses a recipe for a 48-fluid ounce punch that calls for 23 fluid ounces of fruit juice and liquor. The rest is club soda. How much of the recipe is club soda?

STEP 1 *question:* How much of the recipe is club soda?

STEP 2 *necessary information:* 48-fluid ounce punch, 23 fluid ounces of fruit juice and liquor

STEP 3 Draw a diagram, and decide whether to add or subtract.

The diagram shows that you can find the remaining contents by subtraction.

punch − fruit juice and liquor = club soda

23 fl oz juice and liquor
? fl oz club soda
48 fl oz total

STEP 4 Do the arithmetic.

48 fl oz − 23 fl oz = **25 fl oz**

STEP 5 Make sure your answer is sensible. It makes sense that the number of ounces of club soda is less than the number of ounces of punch.

EXAMPLE 2 After losing $237 at the blackjack table, Yolanda had $63 left to spend for the rest of her vacation. How much spending money did she bring?

STEP 1 *question:* How much spending money did she bring?

STEP 2 *necessary information:* $237 lost, $63 left

STEP 3 Draw a diagram, and decide whether to add or subtract.

The diagram shows that you can find the total spending money by addition.
$ lost + $ left = $ total spending money

$237 lost

$63 left

$? total brought on trip

STEP 4 Do the arithmetic.

$237 + $63 = **$300**

STEP 5 Make sure your answer is sensible. Since she lost money, it makes sense that Yolanda started with more money than she has now.

EXAMPLE 3   The oil tanker *Whyon* was loaded with 150,000 barrels of crude oil when it struck a reef and spilled most of its oil. Within a week, the cleanup crew had pumped all the remaining oil from the tanker. If the cleanup crew pumped 97,416 barrels of oil from the ship, how many barrels of oil were spilled?

STEP 1   *question:* How many barrels of oil were spilled?

STEP 2   *necessary information:* 150,000 barrels loaded, 97,416 barrels pumped

STEP 3   Draw a diagram, and decide whether to add or subtract.

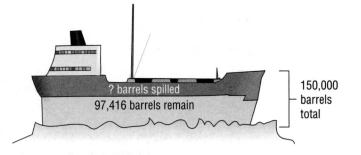

The diagram shows that you can find the amount spilled by subtraction.

total load − amount left = amount spilled

STEP 4   Do the arithmetic.

150,000 barrels
− 97,416 barrels
52,584 barrels

STEP 5   Make sure your answer is sensible. It makes sense that the amount spilled is less than the total amount of oil that was originally on the tanker.

For each problem, make a drawing or a diagram and decide whether to add or subtract. Then solve the problem.

1. If 3 more workers are added to this production team, there will be 31 workers. How many workers are on the production team now?

2. Rafael Hernandez paid $39 less in taxes last year than this year. He paid $483 last year. How much did he pay this year?

3. Every hour 12,000 gallons of water flow through the dam spillway. The 41-year-old dam operator plans to decrease the flow by 3,500 gallons an hour. What will be the new rate of water flow?

4. A $120 ink-jet printer costs $359 less than a laser printer. How much does the laser printer cost?

5. An ink-jet printer costs $120 less than a $359 laser printer. How much does the ink-jet printer cost?

6. Between 6 P.M. and 11 P.M., the temperature decreased by 13 degrees to 61 degrees. What was the temperature at 6 P.M.?

7. At 6 P.M. the temperature was 61 degrees. Between 6 P.M. and 11 P.M., it decreased by 13 degrees. What was the temperature at 11 P.M.?

8. Sixty years ago, 1,412 people graduated from Lincoln High School. Today, 457 of these graduates are still living. How many of the graduates have died?

9. Marion receives an interest-free student loan for $6,000. She pays back $3,800. How much does she still owe?

10. Suppose a 2,600-pound truck can carry a 1,000-pound load. How much does the fully loaded truck weigh?

11. Gamma Airlines allows each passenger to check a maximum of two bags that together must weigh less than 70 pounds. At check-in, Khanh's bags weighed 24 pounds and 42 pounds. Did the total weight of her bags exceed the weight limit?

12. MacroGames Corporation sells 790,000 copies of their video game Fog. They estimate there are 600,000 illegal copies of Fog in circulation. If their estimate is correct, how many copies of Fog are in circulation?

Using Number Sentences to Solve Word Problems

Addition and subtraction word problems can be solved by writing number sentences. A **number sentence** restates a word problem first in words and then in numbers.

EXAMPLE 1 Lori went to school for 5 years in Levittown before moving to Plainview. She then went to school for 7 years in Plainview. For how many years did she go to school?

To write a number sentence, first write the information in the problem in words.

Levittown years plus Plainview years equals total years

Then substitute numbers and mathematical symbols for the words.

5 years + 7 years = total years

Solve.

$$\begin{array}{r} 5 \\ +7 \\ \hline 12 \end{array}$$

12 years = total years

EXAMPLE 2 A play ran for two nights at a theater with a seating capacity of 270 people. The first night 235 people saw the play, and 261 people saw the play the second night. How many people saw the play during its two-night run?

STEP 1 *question:* How many people saw the play during its two-night run?

STEP 2 *necessary information:* 235 people, 261 people

STEP 3 *number sentence:*

first night + second night = total people

235 people + 261 people = total people

$$\begin{array}{r} 235 \\ +261 \\ \hline 496 \end{array}$$

STEP 4 **496 people** = total people

EXAMPLE 3 Gloria bought a $97 dress on sale for $29. How much did she save?

STEP 1 *question:* How much did she save?

STEP 2 *necessary information:* $97, $29

STEP 3 *number sentence:*

original price − sale price = savings

$97 − $29 = savings

$$\begin{array}{r} 97 \\ -29 \\ \hline 68 \end{array}$$

STEP 4 **$68** = savings

Underline the necessary information. Write a word sentence and a number sentence. Then solve the problem.

1. Ross needed a 44-cent stamp. If he paid for the stamp with a half dollar, how much change did he get?

2. John drives 32 miles to work each day. When he arrived at work on Monday, he found that he had already driven 51 miles that day. How many additional miles did John drive on Monday?

3. The theater company needs to sell 172 Saturday tickets to break even. How many more Saturday tickets must they sell in order to break even according to the chart at the right?

Ticket Sales	
Thursday	120
Friday	145
Saturday	134

4. Wendy decides to buy a pre-owned car for $6,300. She has $1,460 saved. She got a loan for the rest. What was the amount of the loan?

5. Barbara Sinclair needs 150 names on her nomination petition to run for office. She collected 119 names on her first day of campaigning. How many more names does she need to collect?

6. After losing 47 pounds, Anna weighed 119 pounds. What was her original weight?

7. A jacket on sale was reduced by $13 to $168. How much was the jacket before the reduction?

8. The refrigerator shown at the right was marked down to $679. How much would a consumer save by buying the refrigerator on sale?

orig. $865

9. John has $2,130 withheld for federal income tax. In fact, he only owes $1,850. How much of a refund does he receive?

10. A car factory cut production by 3,500 cars to 8,200 cars a month. What was the monthly production before the cutback?

11. Closeout Shoes spent $208,682 last month for shoes that were later sold for a total of $327,991. Not counting other expenses, how much profit did the store make on the shoes?

12. Mr. Crockett's cow, Bertha, produced 1,423 gallons of milk last year. His other cow, Calico, produced 1,289 gallons. How much milk did his cows produce last year?

13. Memorial Stadium has a seating capacity of 72,070. The attendance for tonight's game is 58,682. How many seats are empty?

14. In one section of his truck garden, a farmer grew spinach. When the spinach was harvested, he grew green beans. The spinach was harvested after 49 days. The green beans were harvested after 56 days. For how many days were vegetables growing in the truck garden?

ADDITION AND SUBTRACTION WORD PROBLEMS: DECIMALS AND FRACTIONS

Using the Substitution Method

So far, you have solved addition and subtraction word problems using whole numbers. However, many students worry when they see word problems using large whole numbers, fractions, or decimals.

Read the following examples and think about their differences and similarities.

EXAMPLE 1   A manufacturer makes cardboard that is 4 millimeters thick. To save money, he plans to make cardboard that is 3 millimeters thick instead. How much thinner is the new cardboard?

STEP 1   *question:* How much thinner is the new cardboard?

STEP 2   *necessary information:* 4 mm, 3 mm

$$\begin{array}{r} 4 \\ -\ 3 \\ \hline 1 \end{array}$$

STEP 3   Decide what arithmetic operation to use. You are given the thickness of each piece of cardboard. Since you must find the difference between the two pieces, you should subtract.

STEP 4   4 mm − 3 mm = **1 mm**

Note: *mm* stands for *millimeter*. You should be able to do this type of problem even if you aren't familiar with the units of measurement.

EXAMPLE 2   A manufacturer makes cardboard that is 6.45 millimeters thick. To save money, he plans to make cardboard that is 5.5 millimeters thick instead. How much thinner is the new cardboard?

STEP 1   *question:* How much thinner is the new cardboard?

STEP 2   *necessary information:* 6.45 mm, 5.5 mm

$$\begin{array}{r} 6.45 \\ -\ 5.50 \\ \hline 0.95 \end{array}$$

STEP 3   Decide what arithmetic operation to use. You are given the thickness of each piece of cardboard. Since you must find the difference between the two pieces, you should subtract. Be sure to put the decimal points one under the other.

STEP 4   6.45 mm − 5.50 mm = **0.95 mm**

EXAMPLE 3 A manufacturer makes cardboard that is $\frac{3}{8}$ inch thick. To save money, he plans to make cardboard that is $\frac{1}{3}$ inch thick instead. How much thinner is the new cardboard?

STEP 1 *question:* How much thinner is the new cardboard?

STEP 2 *necessary information:* $\frac{3}{8}$ inch, $\frac{1}{3}$ inch

STEP 3 Decide what arithmetic operation to use. You are given the thickness of each piece of cardboard. Since you must find the difference between the two pieces, you should subtract.

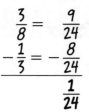

$$\begin{aligned} \frac{3}{8} &= \frac{9}{24} \\ -\frac{1}{3} &= -\frac{8}{24} \\ \hline &\frac{1}{24} \end{aligned}$$

STEP 4 $\frac{3}{8}$ in. $- \frac{1}{3}$ in. $= \frac{1}{24}$ **in.**

Note: Remember, to find the solution, you must change unlike fractions to fractions with a common denominator.

What did you notice about the three example problems?

The wording of all three is exactly the same. Only the numbers and labels have been changed. All three problems are solved the same way, by subtracting.

Then, why do Examples 2 and 3 seem harder than the first example?

The difficulty has to do with **math intuition,** or the feel that a person has for numbers. You have a very clear idea of the correct answer to $4 - 3$. It is more difficult to picture $7,483,251 + 29,983$ or $6.45 - 5.5$. And for most of us, our intuition totally breaks down for $\frac{3}{8} - \frac{1}{3}$.

Changing only the numbers in a word problem does not change what must be done to solve the problem. By substituting small whole numbers in a problem, you can understand the problem and how to solve it.

Fractions, especially those with different denominators, are especially hard to picture. You can make the problem easier to understand by substituting small whole numbers for the fractions. You can substitute any numbers, but try to use numbers under 10. These numbers do not have to look like the numbers they are replacing.

Once you make your decision about *how* to solve the problem, you can return the original numbers to the word problem and work out the solution.

Below is a set of six substitutions. Each of the substitutions will fit only one of the following six word problems. Match the letter of the correct substitution to each problem. (After each problem, you are told which numbers to substitute for that problem. To keep it simple, we are only using the numbers 4, 3, and 1 in the substitutions.)

> **a.** 3 pounds + 4 pounds = 7 pounds
> **b.** 4 pounds − 3 pounds = 1 pound
> **c.** $3 − $1 = $2
> **d.** $3 + $1 = $4
> **e.** 3 inches + 1 inch = 4 inches
> **f.** 3 inches − 1 inch = 2 inches

1. A sweater that normally sells for $35.99 has been marked down by $10.99. What is the sale price of the sweater?

   Substitute $3 for $35.99 and $1 for $10.99.

   _____

2. How much heavier is the roast on the right than the roast on the left?

   Substitute 3 for 3.46 and 4 for 4.17.

   _____

3.46 lb   4.17 lb

3. Robin bought a 3.28-pound steak and a 4.84-pound chicken. What was the weight of the meat she bought?

   Substitute 3 for 3.28 and 4 for 4.84.

   _____

4. Isatu bought a skirt for $31.99 and a top for $11.59. How much did she spend in all?

   Substitute $3 for $31.99 and $1 for $11.59.

   _____

5. Michael caught a $21\frac{1}{4}$-inch fish. His friend Paul caught a $23\frac{1}{16}$-inch fish. How much longer was Paul's fish?

   Substitute 1 for $21\frac{1}{4}$ and 3 for $23\frac{1}{16}$.

   _____

6. Two boards were placed end-to-end. The first board was $40\frac{7}{8}$ inches long. The second board was $32\frac{3}{4}$ inches long. What was the combined length of the 2 boards?

   Substitute 3 for $40\frac{7}{8}$ and 1 for $32\frac{3}{4}$.

   _____

Using Estimation

When your car is in an accident and you take it to an auto body shop for repairs, you first receive an **estimate** for the cost of the repairs. This might not be the exact or final price, but it should be close.

When solving word problems, it is also important to have some idea of what the answer should be before you start doing the arithmetic. You can get an estimate of the answer by approximating the numbers in the problem.

An **estimate** is almost, but not quite, the exact number. For instance,

> In the last election, the newspaper reported that Alderman Jones received 52% of the vote and his opponent received 48%. Actually, the alderman received 52.1645% of the vote and his opponent received 47.8355%.

The newspaper did not report the exact percent of the vote; it **rounded** the numbers to the nearest whole percent. Rounded numbers are one type of estimate.

Estimating the numbers and doing quick arithmetic in your head is a good way to check your work. Throughout the rest of this workbook, you should first estimate a solution before doing the calculation.

Below is a set of six estimated solutions. The numbers in the solutions have been rounded. Match each word problem with one of the solutions.

> **a.** 9 miles − 7 miles = 2 miles
> **b.** 9 miles + 7 miles = 16 miles
> **c.** 4 billion dollars − 1 billion dollars = 3 billion dollars
> **d.** 4 billion dollars + 1 billion dollars = 5 billion dollars
> **e.** 37,000 fans − 35,000 fans = 2,000 fans
> **f.** 37,000 fans + 35,000 fans = 72,000 fans

1. During the last weekend in July, 35,142 fans saw a baseball game on Saturday. On Sunday, 36,994 fans saw a game. What was the total attendance for the weekend?

   _____

2. Nationwide, Grand Discount stores sold 37,238 window fans in April and 34,982 fans in May. How many more fans were sold in April?

   _____

3. The original estimate for the cost of a nuclear power plant was 0.984 billion dollar. The final cost was 4.16 billion dollars. How much did the price increase from the original estimate?

   _____

4. The city budget is 3.92 billion dollars. It is expected to increase 1.2 billion dollars over the next 5 years. How much is the budget expected to be 5 years from now?

   _____

5. By expressway, it is $7\frac{1}{4}$ miles to the beach. By backroads, it is $8\frac{9}{10}$ miles. How much shorter is the trip when driving by expressway?

   _____

6. Pat is a long-distance runner. He ran $6\frac{9}{10}$ miles on Saturday and $9\frac{1}{8}$ miles on Sunday. How many miles in all did he run during the weekend?

   _____

Decimals: Restating the Problem

Restating the problem is one method that will work as well with solving decimal problems as with whole-number problems. Don't worry about the decimal points until after you have decided to add or subtract. Then remember to line up the decimal points before doing the arithmetic.

EXAMPLE A pair of pants was on sale for $18.99. A shirt was on sale for $16.49. Alan decided to buy both. How much did he spend?

STEP 1 *question:* How much did he spend?

STEP 2 *necessary information:* $18.99, $16.49

STEP 3 *restatement:* Since Alan is buying both items, you add to find the total amount he spent.

$$\begin{array}{r} 18.99 \\ + 16.49 \\ \hline 35.48 \end{array}$$

STEP 4 $18.99 + $16.49 = **$35.48**

STEP 5 Round $18.99 up to $19, and round $16.49 down to $16.

$19 + $16 = $35

Therefore, your answer should be close to $35. Making an estimate is a good method to check your answer and to make sure it is sensible.

Circle the letter of the correct restatement and solve the problem. Use estimation to make sure your answer is sensible.

1. Using his odometer, George discovered that one route to work was 6.3 miles long and the other was 7.1 miles. How much shorter was the first route?

   a. Since you are given the two distances to work, add to find out how much shorter was the first route.

   b. To find how much shorter was the first route, subtract to find the difference.

2. Max had to put gasoline in his 8-year-old car twice last week. The first time, he put in 9.4 gallons. The second time, he put in 14.7 gallons. How much gasoline did he put in his car last week?

    a. To find the total amount of gasoline, you add.

    b. Since you are given the two amounts of gasoline, you subtract to find the difference.

3. The first fish fillet weighed 1.42 pounds. The second fillet weighed 0.98 pound. Alicia decided to buy both fillets. What was the weight of the fish she bought?

    a. To find the total weight of the two fish fillets, you add.

    b. Since you are given the weight of the two fish fillets, you subtract to find the difference between their weights.

4. At the Reckless Speedway, Bobby was clocked at 198.7 mph, while Mario was clocked at 200.15 mph. How much faster did Mario drive than Bobby?

    a. Add the two speeds to find Mario's total speed.

    b. Since Mario drove faster, subtract Bobby's speed from his to find out the difference between the speeds.

5. Last year the unemployment rate was 7.9%. This year it has increased to 9.1%. By how much did unemployment rise?

    a. Since you are looking for the total amount by which unemployment rose, add the two unemployment rates.

    b. Since you are looking for the difference between the unemployment rates, subtract last year's rate from this year's rate.

Decimals: Drawings and Diagrams

Diagrams and drawings can help you solve decimal addition or subtraction word problems.

EXAMPLE A metal bearing was 0.24 centimeter thick. The machinist ground it down until it was 0.065 centimeter thinner. How thick was the metal bearing after it had been ground down?

0.24 cm thick

0.065 cm ground down

new thickness

STEP 1 *question:* How thick was the metal bearing after it had been ground down?

STEP 2 *necessary information:* 0.24 cm, 0.065 cm

STEP 3 *make a drawing:* To find the size of the bearing after it was ground down, you subtract.

STEP 4 Do the arithmetic. Be sure to line up the decimal points. If you add a zero, you can see that 0.24 (0.240) is greater than 0.065.

$$\begin{array}{r} 0.240 \\ -\ 0.065 \\ \hline 0.175 \end{array}$$

0.240 cm − 0.065 cm = **0.175 cm**

Remember: When subtracting decimals, first line up the decimal points. Then fill any blank spaces to the right of the decimal point with zeros. This should help you borrow correctly.

Make a drawing or a diagram, and solve the problem. (Each person's drawing may be different. What is important is that the diagram makes sense to you.)

1. Meatball subs used to cost $4.50 at Mike's, but he just raised the price by $0.45. How much do meatball subs cost now?

2. Tara's prescription for 0.55 gram of antibiotic was not strong enough. Her doctor gave her a new prescription for 0.7 gram of antibiotic. How much stronger was the new prescription?

3. Mike Johnson was hitting .342 before he went into a batting slump. By the end of his slump, his average had dropped .083. What was his batting average at the end of his slump?

4. Kaiya earned $713.50 and had $226.13 taken out for deductions. How much was her take-home pay?

5. A wooden peg is 1.6 inches wide and 3.2 inches long. It can be squeezed into an opening 0.05-inch smaller than the width of the peg. What is the width of the smallest possible opening into which the peg could be squeezed?

6. By midweek Commonwealth Motors had spent $46.65 on snacks and coffee for its customers. At the end of the week, they had spent $63.35 more. How much did Commonwealth Motors spend that week on snacks and coffee for its customers?

7. The gap of a spark plug should be 0.0038 inch. The plug would still work if the gap were off by as much as 0.0002 inch. What is the largest gap that would still work?

8. The King Coal Company mined 126.4 tons of coal. Because of high sulfur content, 18.64 tons of coal were unusable. How many tons of coal were usable?

9. Anya was mixing chemicals in a lab. The formula called for 1.45 milliliters of sulfuric acid, but she had 1.8 milliliters of sulfuric acid in her pipette. How much extra sulfuric acid does she have in the pipette? (A pipette is a glass tube used for measuring chemicals.)

10. Barbara complained that a 2.64-pound steak had too much excess fat and bone. The butcher trimmed the steak and reweighed it. It then weighed 2.1 pounds. How many pounds of fat and bone did the butcher cut off the steak?

Decimals: Writing Number Sentences

Number sentences can help you solve decimal addition and subtraction word problems. Look at the following examples to see how number sentences are used.

EXAMPLE 1   Meryl bought $66.27 worth of groceries and paid with a $100 bill. How much change did she receive?

STEP 1   *question:* How much change did she receive?

STEP 2   *necessary information:* $66.27, $100

STEP 3   *number sentence:*
amount paid − price of groceries = change
$100.00 − $66.27 = change

$$\begin{array}{r} 100.00 \\ -\ 66.27 \\ \hline 33.73 \end{array}$$

STEP 4   **$33.73** = change

EXAMPLE 2   Store manager Dioreles Garcia marked down a pair of pants to $32.49. She had discounted the pants $24.49 from the original price. What was the original price?

STEP 1   *question:* What was the original price?

STEP 2   *necessary information:* $32.49, $24.49

STEP 3   *number sentence:*
sale price + discount = original price
$32.49 + $24.49 = original price

$$\begin{array}{r} 32.49 \\ +24.49 \\ \hline 56.98 \end{array}$$

STEP 4   **$56.98** = original price

Underline the necessary information. Write a word sentence and a number sentence. Then solve the problem.

1. The Sticky Candy Company decided to reduce the size of their chocolate candy bar by 0.6 ounce to 2.4 ounces. How much did the chocolate bar weigh before the change?

2. Julia's lunch cost $7.38. If she paid with a $10 bill, how much change did she get?

3. Rosa bought one chicken that weighed 3.94 pounds and one that weighed 4.68 pounds. She also bought a 1.32-pound steak. How much chicken did she buy?

4. The odometer at the right shows Steve's car mileage when he left Boston. When he arrived in New York, the odometer read 23,391.4 miles. How long was the trip?

2	3	1	7	2	.	3

5. The chart at the right shows the costs of subway and bus rides in Connie's city. If Connie needs to take one bus and one subway to her mother's house, how much will it cost her for a one-way trip?

Fares	
Bus	$1.40
Subway	$1.65

6. Judy spent $634.73 for a new washing machine in Massachusetts. If she bought the same machine in New Hampshire, she would have paid $595.99, since that state does not have a sales tax. How much less would she have paid in New Hampshire?

7. When he ran the 200-meter race, Marcus ran the first 100 meters in 14.36 seconds and the second 100 meters in 13.9 seconds. What was his total time for the race?

8. One assembly line at the plant produced 966 soda bottles in one hour. Another line produced 50 fewer bottles in the same amount of time. How many bottles did the second line produce in an hour?

Fractions: Restating the Problem

In this section, you will restate the problem in order to decide whether to add or subtract. Before solving these fraction problems, you might want to use estimation to help you decide what arithmetic operation to use.

EXAMPLE Tanya grew $2\frac{3}{4}$ inches last year. If she was $42\frac{1}{2}$ inches tall a year ago, how tall is she now?

STEP 1 *question:* How tall is she now?

STEP 2 *necessary information:* $2\frac{3}{4}$ inches, $42\frac{1}{2}$ inches

STEP 3 *restatement:* Since you know Tanya's old height, and you know that she grew, you must add to find her new height.

STEP 4 *estimation:* 3 inches + 43 inches = 46 inches

STEP 5 $2\frac{3}{4}$ inches + $42\frac{1}{2}$ inches = height now

$2\frac{3}{4} + 42\frac{2}{4} = 44\frac{5}{4}$ inches = **$45\frac{1}{4}$ inches now**

$$\begin{array}{rcl} 2\frac{3}{4} &=& 2\frac{3}{4} \\ + 42\frac{1}{2} &=& + 42\frac{2}{4} \\ \hline & & 44\frac{5}{4} = 45\frac{1}{4} \end{array}$$

Remember: Whenever you add or subtract fractions, find a common denominator.

Each problem is followed by two restatements and estimations.
Circle the correct restatement. Then solve the problem.

1. A carpenter needed one piece of molding $28\frac{1}{2}$ inches long and a second piece $31\frac{1}{4}$ inches long. How much molding did he need?

   a. To find the difference between the pieces of molding needed, you should subtract.
   31 inches − 29 inches = 2 inches

   b. To find the total amount of molding needed, you should add.
   29 inches + 31 inches = 60 inches

2. Aldo combined $1\frac{2}{3}$ cups of flour and $1\frac{1}{3}$ cups of butter in a 2-quart mixing bowl. How many cups of the mixture did Aldo have?

    a. Since Aldo is combining the flour and butter, the amount of the mixture can be found by adding.

       2 cups $+$ 1 cup $=$ 3 cups

    b. Since you are given the amount of flour and the amount of butter, you subtract to find the amount of the mixture.

       2 cups $-$ 1 cup $=$ 1 cup

3. According to the scales at the right, how much heavier is Tara than Erin?

Tara

    a. Since you are comparing two weights, subtract to find the difference.

       71 pounds $-$ 63 pounds $=$ 8 pounds

    b. Since you are finding their total weight, add the given weights.

       71 pounds $+$ 63 pounds $=$ 134 pounds

Erin

4. Mira was $18\frac{3}{4}$ inches long at birth. Six months later, she was $23\frac{1}{4}$ inches long. How much longer was Mira at 6 months than at birth?

    a. To find how much longer Mira is, add the given lengths.

       19 inches $+$ 23 inches $=$ 42 inches

    b. To find how much longer Mira is, subtract her birth length from her length at 6 months.

       23 inches $-$ 19 inches $=$ 4 inches

Fractions: Diagrams and Pictures

Making diagrams and pictures can also help you solve fraction addition or subtraction word problems.

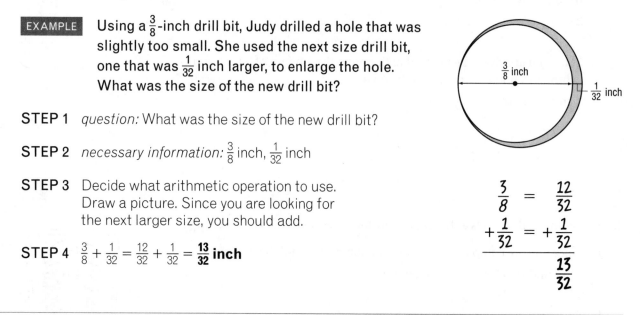

EXAMPLE Using a $\frac{3}{8}$-inch drill bit, Judy drilled a hole that was slightly too small. She used the next size drill bit, one that was $\frac{1}{32}$ inch larger, to enlarge the hole. What was the size of the new drill bit?

STEP 1 *question:* What was the size of the new drill bit?

STEP 2 *necessary information:* $\frac{3}{8}$ inch, $\frac{1}{32}$ inch

STEP 3 Decide what arithmetic operation to use. Draw a picture. Since you are looking for the next larger size, you should add.

STEP 4 $\frac{3}{8} + \frac{1}{32} = \frac{12}{32} + \frac{1}{32} = \frac{13}{32}$ **inch**

$$\frac{3}{8} = \frac{12}{32}$$
$$+\frac{1}{32} = +\frac{1}{32}$$
$$\overline{\qquad \frac{13}{32}}$$

Make a drawing or diagram and solve each of the problems below.

1. A 2-by-4 (a 2-inch by 4-inch board) is really not 4 inches wide. It is $\frac{5}{8}$ inch narrower. What is the real width of the board?

2. Yiu is at the hospital for a total of $8\frac{1}{2}$ hours a day. If during each day he has $1\frac{3}{4}$ hours for breaks, how long does he work each day?

3. Hope worked $6\frac{1}{2}$ hours and took an additional $\frac{3}{4}$ hour for lunch. What was the total amount of time that Hope spent at work and lunch?

4. A 2-by-4 is not really 2 inches thick. It is $\frac{1}{2}$ inch thinner. What is the real thickness of the board?

5. A recipe called for $2\frac{1}{2}$ cups of flour. George only had $1\frac{2}{3}$ cups. How much flour did he need to borrow from his neighbor?

6. Felix tried to loosen a bolt with a $\frac{3}{4}$-inch wrench, but it was too large. He decided to try the next smaller size, which was $\frac{1}{16}$ inch smaller. What was the size of the next smaller size wrench?

Fractions: Using Number Sentences

Number sentences can also help you solve fraction addition or subtraction word problems.

EXAMPLE   David plans to make a 3-inch-thick insulated roof. The roof will be made with a layer of thermal board on top of $\frac{5}{8}$-inch plywood. How thick can the thermal board be?

STEP 1   *question:* How thick can the thermal board be?

STEP 2   *necessary information:* 3 inches, $\frac{5}{8}$ inch

STEP 3   *number sentence:*
thickness of roof − plywood = thermal board
3 inches − $\frac{5}{8}$ inch = thermal board

$$\begin{array}{r} 3 = 2\frac{8}{8} \\ -\frac{5}{8} = -\frac{5}{8} \\ \hline 2\frac{3}{8} \end{array}$$

STEP 4 $2\frac{3}{8}$ **inches** = thickness of thermal board

---

Write a word sentence and a number sentence for each problem. Then solve the word problems.

1. Elina filled a 3-quart punch bowl. If she used $1\frac{1}{4}$ quarts of rum, how many quarts of other liquid ingredients did she use to fill the bowl?

2. Amy bought a skirt that was the length shown below. If she shortened it to $32\frac{3}{4}$ inches, how much did she take off?

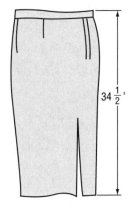

$34\frac{1}{2}$'

3. After 3 weeks in the store, a bolt of cloth that had originally been 20 yards long was $6\frac{1}{2}$ yards long. Then, $3\frac{2}{3}$ more yards of the cloth were sold. How much cloth was left?

4. Last winter Sergio used $\frac{1}{8}$ cord of wood one week and $\frac{1}{12}$ cord of wood the next week to heat his house. How much wood did he use during the 2 weeks?

5. Lydia bought $62\frac{1}{2}$ inches of cloth to make drapes. She used $\frac{3}{4}$ inch for the hem. How long were the drapes?

Using Algebra to Solve Word Problems

You have already seen how to use number sentences to solve addition and subtraction word problems. In algebra, instead of using words, you use a letter of the alphabet to stand for the number you are looking for.

A **number sentence** in which one amount is equal to another amount is called an **equation**. To find what number will make an equation true, you must solve the equation. One way to solve an equation is shown in the following example.

EXAMPLE A cup of 2% milk contains 130 calories, of which 45 calories are from fat. How many calories in the cup of 2% milk are not from fat?

STEP 1 *question:* How many calories in the cup of 2% milk are not from fat?

STEP 2 *necessary information:* 130 calories, 45 calories

STEP 3 *number sentence or equation:*

calories from fat + calories not from fat = total calories

45 calories + calories not from fat = 130 calories

You can rewrite the number sentence using the letter c to stand for *calories not from fat*. (You can select any letter, but c is a good choice since it is the first letter of *calories*.) 45 calories + c = 130 calories

STEP 4 *solve the equation:*

One way to solve the equation is to subtract 45 calories from each side so that c stands alone on the left side of the equation. You can do this because if you subtract the same amount from both sides of an equation, the results are still equal.

$$
\begin{array}{rcr}
45 \text{ calories} + c &=& 130 \text{ calories} \\
-\ 45 \text{ calories} &=& -\ 45 \text{ calories} \\
\hline
c &=& 85 \text{ calories}
\end{array}
$$

STEP 5 *Does the answer make sense?*

45 calories + 85 calories = 130 calories. The answer makes sense.

Estimation: The letter is a placeholder for a number. You need to find the number that will make the equation true. To build your math intuition, try to substitute a number for the letter before you start the formal solution. Try to get a good estimate of the answer before doing the detailed calculations.

Write a word sentence and an equation for each problem. Then solve the word problem.

1. In January the adult education student Website on the Internet had 2,917 hits or visits. On February the site had 4,348 hits. How many more hits did the site have in February?

2. Belquis used her debit card to buy $78.62 of groceries. Afterwards she checked her balance and found that she had $138.79 left in the account. How much was in her account before she paid for her groceries?

3. A slice of cheese pizza has 251 calories. A slice of pepperoni pizza has 290 calories. How many calories are in the pepperoni on a slice of pepperoni pizza?

4. In order to patch a water-damaged post, Raymonde needed a $1\frac{3}{4}$ foot long 1" × 6" board. In her basement, she found a $3\frac{1}{2}$ foot long 1" × 6" board. After she cut the board to make the repair, how much was left over?

5. A tablet of Excedrin contains 250 milligrams of acetaminophen, 250 milligrams of aspirin, and 65 milligrams of caffeine as active ingredients. How many milligrams of active ingredients are in a tablet of Excedrin?

6. In the fleece mill, inspector Diogenes found that the pile height of a batch of fabric was $\frac{5}{8}$ inch. The standard pile height for the fabric was $\frac{7}{16}$ inch. How much material needs to be sheared off for the fabric to meet the standard?

Solving Addition and Subtraction Word Problems

Solve each problem and circle the letter of the correct answer.

1. The Hammerhead Nail Company produces 55,572 nails and 4,186 screws a day. On Monday 1,263 nails were no good. How many good nails were made on Monday?

   a. 56,835 nails

   b. 54,309 nails

   c. 59,758 nails

   d. 51,386 nails

   e. 58,495 nails

2. Ana Marie started off with $75.62 in her wallet. How much money did she have left after spending $38.56?

   a. $114.18

   b. $37.06

   c. $1.98

   d. $71.76

   e. none of the above

3. The gauges at the right show Ron's mileage using different types of fuel. How much better was his mileage when he used biodiesel?

   a. 3.1 miles

   b. 55.3 miles

   c. 43.1 miles per gallon

   d. 2.9 miles per gallon

   e. 15.3 miles per gallon

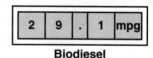

4. When Kathy bought her car, she paid $1,800 down and had $640 left in her savings account. She then paid $6,400 over the next 2 years to finish paying for the car. How much did the car cost her?

   a. $8,840

   b. $4,600

   c. $4,100

   d. $8,200

   e. $7,560

5. Before cooking, a hamburger weighed $\frac{1}{4}$ pound. After cooking, it weighed $\frac{3}{16}$ pound. The rest of the hamburger was fat that burned off during cooking. How much fat burned off during cooking?

    a. $\frac{4}{20}$ pound

    b. $\frac{1}{6}$ pound

    c. $\frac{1}{16}$ pound

    d. $\frac{7}{16}$ pound

    e. none of the above

6. Brand X contains 0.47 gram of pain reliever per 1.5-gram tablet. Brand Y contains 0.6 gram of pain reliever. How much more pain reliever does Brand Y have than Brand X?

    a. 0.53 gram

    b. 0.41 gram

    c. 0.13 gram

    d. 0.27 gram

    e. 2.57 grams

7. A public television station has already raised $391,445 and must raise $528,555 more to stay in business. What was the target amount for the station's fund-raising drive?

    a. $920,000

    b. $137,110

    c. $127,110

    d. $237,110

    e. none of the above

8. Nam bought the window shade at the right. When he got home, he found that it was $2\frac{3}{8}$ inches too narrow. What was the width he needed?

    a. $24\frac{3}{8}$ inches

    b. $29\frac{1}{8}$ inches

    c. $28\frac{1}{2}$ inches

    d. 24 inches

    e. none of the above

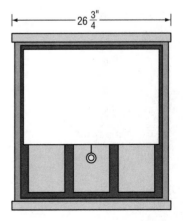

$26\frac{3}{4}$"

9. To heat her house last winter, Mrs. Garcia used $\frac{5}{8}$ cord of wood in February and $\frac{1}{12}$ cord of wood in March. How much wood did she use?

   a. $\frac{3}{10}$ cord

   b. $\frac{13}{24}$ cord

   c. $\frac{1}{2}$ cord

   d. $\frac{3}{4}$ cord

   e. none of the above

10. A gallon of economy paint contained 3.4 tubes of pigment per gallon. The high-quality paint contained 5.15 tubes of pigment per gallon. What was the difference between the amount of pigment used for each paint?

   a. 4.81 tubes

   b. 5.49 tubes

   c. 8.55 tubes

   d. 2.35 tubes

   e. none of the above

11. A radioactive tracer lost $\frac{1}{2}$ of its radioactivity in an hour. Three hours later it had lost another $\frac{7}{16}$ of its radioactivity. What was the total loss in radioactivity for the entire time?

   a. $\frac{7}{32}$ of its radioactivity

   b. $\frac{15}{16}$ of its radioactivity

   c. $\frac{4}{9}$ of its radioactivity

   d. $\frac{1}{16}$ of its radioactivity

   e. none of the above

12. A gypsy moth grew 0.03 gram from the size shown at the right. How much did the moth weigh after it grew?

   a. 2.08 grams

   b. 2.74 grams

   c. 3.10 grams

   d. 2.80 grams

   e. none of the above

Gypsy Moth,
2.77 grams

MULTIPLICATION AND DIVISION WORD PROBLEMS: WHOLE NUMBERS

Identifying Multiplication Key Words

In arithmetic, there are four basic operations: addition, subtraction, multiplication, and division. As you have seen, subtraction can be thought of as the opposite of addition. In the same way, division can be thought of as the opposite of multiplication. This concept is useful in deciding whether a problem is a multiplication or a division problem.

In previous chapters, you looked at addition and subtraction key words. There are also multiplication key words.

EXAMPLE 1   Diane always bets $2 on a race. Last night she bet eight times. How much money did she bet?

multiplication key word: times

EXAMPLE 2   It cost Fernando $39 per day to rent a car. He rented a car for 4 days. How much did he pay to rent the car?

multiplication key word: per

Remember: *Per means* "for each."

Multiplication can also be considered repeated addition. Therefore, it is possible for an addition key word to also be a multiplication key word. *Total* is a word that can indicate either addition or multiplication.

In the following problems, circle the multiplication key words. DO NOT SOLVE!

1. Miguel pays his landlord $870 for rent 12 times a year. How much rent does he pay in a year?

2. During the 9 months that she stayed in her apartment, Isabelle paid $83 per month for electricity. How much did she pay for electricity during the time she stayed in her apartment?

3. At the stable, one horse eats 3 pounds of hay a day. What is the total amount of hay needed to feed 26 horses?

4. When her children were young, Lizette had a part-time job for 18 hours a week. She now works twice as many hours as she did then. How many hours a week does she work now?

5. Sam and Marion bought a new home on an 80-by-90 foot lot. How large, in square feet, was the lot?

6. Amy's living room is 24 feet long and 16 feet wide. What is the area of her living room?

7. Sally's research found that every dollar invested in campaign fund-raising was multiplied eight times by new contributions. If her research is correct, how much in new contributions can she expect if she invests $3,600 in fundraising?

8. Fresh orange juice has 14 calories per fluid ounce. How many calories are in an 8-fluid ounce serving of orange juice?

9. Margo's family drinks 3 gallons of milk per week. At $3.79 per gallon, how much does Margo spend each week for milk?

Solving Multiplication Word Problems with Key Words

Look at the following examples of multiplication key words.

EXAMPLE 1   Shirley cleans the kitchen sink three times a week. How many times does she clean the sink in 4 weeks?

STEP 1   *question:* How many times does she clean the sink?

STEP 2   *necessary information:* 3 times a week, 4 weeks

$$\begin{array}{r} 3 \\ \times\ 4 \\ \hline 12 \end{array}$$

STEP 3   Decide what arithmetic operation to use.

multiplication key word: times

STEP 4   3 times a week × 4 weeks = **12 times**

EXAMPLE 2   During the Great Depression, eggs cost 14 cents per dozen. How much did 5 dozen eggs cost?

STEP 1   *question:* How much did 5 dozen eggs cost?

STEP 2   *necessary information:* 14 cents per dozen, 5 dozen

$$\begin{array}{r} 14 \\ \times\ 5 \\ \hline 70 \text{ cents} \end{array}$$

STEP 3   Decide what arithmetic operation to use.

multiplication key word: per

STEP 4   14 cents per dozen × 5 = **70 cents**

In the problems below, underline the necessary information, and circle the multiplication key words. Then solve the problem.

1. In order to cover expenses for her small clothing store and make enough money to support herself, Margarita must charge twice as much for a garment than what she paid. If she paid $58 a dress for a shipment of dresses, what must she charge for a dress?

2. Alan needs to buy 4 sets of guitar strings. There are 6 strings per set. How many strings will he buy?

3. Harvey Furniture Company's advertisement plays on the radio five times a day and appears in 12 newspapers. How many times does its ad play on the radio in a week?

4. The We-Fix-it Company charged $75 per hour to repair computers. The We-Fix-it employee worked for 3 hours updating computers at the Long Distance Trucking Company. How much did the Long Distance Trucking Company pay for this work?

Identifying Division Key Words

As you might suspect by now, there are also division key words.

EXAMPLE 1 Ron and Nancy shared equally the cost of an $86 phone bill. How much did each of them pay?

division key words: shared equally, each

Remember: Any word indicating that something is cut up is a division key word.

EXAMPLE 2 The $9 million lottery prize will be divided equally among the 3 winners. How much money will each winner receive?

division key words: divided equally, each

Each is considered a division key word. It indicates you are given an amount for multiples of the same thing and are looking for the value for just one of them.

Circle the division key words. DO NOT SOLVE!

1. Carlos, Dan, and Juan shared the driving equally when they drove from Chicago to Los Angeles. How much did Carlos drive according to the map below?

2. A bakery produced 6,300 chocolate chip cookies in a day. The cookies were packed in boxes with 36 cookies per box. How many boxes were used that day?

3. Three salesmen sold $2,250 worth of power tools. On the average, how much did each of them sell?

4. It cost $192 to rent the gym for the basketball game. If the 16 players shared the cost equally, how much did each of them pay?

Solving Division Word Problems with Key Words

Knowing the division key words can help you solve division word problems.

EXAMPLE 1   Annual union dues of $948 are divided into 12 equal monthly payments. How much is a monthly payment?

STEP 1   *question:* How much is a monthly payment?

STEP 2   *necessary information:* $948 a year, 12 monthly payments

STEP 3   *division key words:* divided, equal

STEP 4 $948 ÷ 12 monthly payments = **$79**

$$\begin{array}{r} 79 \\ 12\overline{)948} \end{array}$$

EXAMPLE 2   A 48-minute basketball game has four equal periods. How long is each period?

STEP 1   *question:* How long is each period?

STEP 2   *necessary information:* 48 minutes, 4 periods

STEP 3   *division key words:* equal, each

STEP 4   48 minutes ÷ 4 periods = **12 minutes**

$$\begin{array}{r} 12 \\ 4\overline{)48} \end{array}$$

For the problems below, underline the necessary information, and circle the division key words. Then solve the word problem.

1. The chocolate bar at the right was shared equally among four children. How much chocolate did each child receive?

12 oz

2. A 60-minute hockey game is divided into three equal periods. How long is the third period?

3. A washing machine that cost $379 when new, now costs $156 used. It can be paid for in 12 monthly payments. How much is each payment on the used washer?

4. A package of 24 mints costs 96 cents. How much does each mint cost?

5. Raffle tickets cost $3 each. If the prizes are worth $4,629, how many tickets must be sold for the raffle to break even?

Key Word Lists for Multiplication and Division

As with addition and subtraction, you can compile lists of multiplication and division key words.

Generally, in multiplication word problems, you are given one of something and asked to find many. You can often think of these problems as multiplying together the number of pieces and the size of each piece to get a total.

MULTIPLICATION KEY WORDS	
multiplied	as much
times	twice
total	by
of	area
per	volume
product	every
tripled	quadrupled
multiple	

Generally, in division word problems, you are given many things and asked to find one. You can often think of these problems as dividing a total by the number of pieces or the size of each piece.

DIVISION KEY WORDS	
divided (equally)	average
split	every
each	out of
cut	ratio
equal pieces	shared
quotient	half
third	quarter

Remember: Key words are only a clue for solving a problem. Any key word can also appear in word problems needing the opposite operation in order to be solved. You need to read for understanding.

Solving Multiplication and Division Problems with Key Words

In this exercise, some of the word problems have multiplication key words and some have division key words. In each problem, circle the key word(s). Identify the circled word(s) as a multiplication or division key word. Then solve the problem.

1. The pizza shown at the right is to be divided equally among four people. How many pieces will each person get?

2. A small oil exploration company made a profit of $900,000 last year. How much money will each of the fifty investors receive if all of the profits are divided equally among them?

3. A gas station owner charged $35 per oil change. In one day he did 15 oil changes. What was the total amount of money he received for oil changes?

4. In the discount store, a dress cost $47. In a full-price store, the same dress was quadruple the price. How much did the dress cost at the full-price store?

5. To earn a high school equivalency certificate, a student in Illinois must score 225 points total on five tests, but no less than 40 points on each test. What is the average score on each test that a student needs to get the certificate?

6. Domingo's taxi averages 38 miles per gallon of gasoline. How many miles can he expect to drive on a full 15-gallon tank?

Deciding When to Multiply and When to Divide

Word problems are rarely so simple that you can solve them just by finding key words. You must develop your comprehension of the meaning of word problems. Key words are an aid to that understanding.

In earlier chapters, you learned that the same key word that helped you decide to add in one problem might also appear in a subtraction problem. The same is true with multiplication and division key words.

Don't Despair!

Learning what the key words mean is the first step to understanding word problems.

In the examples that follow, you are given two numbers and are asked to find a third. In each example, you must decide whether to multiply or divide.

In many word problems, the question will ask you to find a total amount, or it will give you a total amount and ask you to find either the number of pieces or an amount for each piece.

- When you are given the number of pieces and the amount for each piece, and you are asked to find the total, you should multiply.

- When you are given the total and the number of pieces, and you are asked to find the amount for each piece, you should divide.

- When you are given the total and the amount for each piece, and you are asked to find the number of pieces, you should divide.

EXAMPLE 1 Each of the city's 24 snowplows can plow 94 miles of road a day. If all snowplows are running, how many miles of road can be plowed by the city in one day?

STEP 1 *question:* How many miles of road can be plowed?

STEP 2 *necessary information:* 24 snowplows, 94 miles of road

STEP 3 Decide what arithmetic operation to use. Use the following method to draw a diagram.

Draw two boxes, and label them *# of pieces* and *amount of each piece.* Put a times sign between the boxes.

To the right of the two boxes, put an equals sign and another box that you should label *total.*

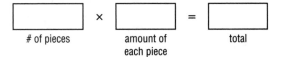

Fill in the boxes with information from the problem. Use only the information that is needed to solve the problem.

Put 24 snowplows in the # of pieces box since that is the number of snowplows. Put 94 miles in the *amount of each piece* box since that is the number of miles each snowplow can plow.

The box that is empty represents the amount that you are looking for, the total number of miles of road.

24 snowplows	×	94 miles	=	
# of pieces		amount of each piece		total

Since the total box is empty, you should multiply to find the total miles.

STEP 4   snowplows × miles per snowplow = total miles

24 snowplows × 94 miles per snowplow = **2,256 miles**

$$\begin{array}{r} 94 \\ \times\ 24 \\ \hline 376 \\ 188 \\ \hline 2,256 \end{array}$$

EXAMPLE 2   A city has 24 snowplows to plow its 2,256 miles of road. How many miles of road must each snowplow cover in order to plow all the roads?

STEP 1   *question:* How many miles of road must each snowplow cover?

STEP 2   *necessary information:* 24 snowplows, 2,256 miles

STEP 3   Draw and label the boxes.

	×		=	
# of pieces		amount of each piece		total

Put 24 snowplows in the *# of pieces* box since that is the number of snowplows. Put 2,256 miles in the *total* box since that has been given as the total number of miles.

24 snowplows	×		=	2,256 miles
# of pieces		amount of each piece		total

The box that is empty represents the amount that you are looking for. Since the total has been given, you should divide to find the amount of each piece, in this case the number of miles plowed by each snowplow.

STEP 4   total miles ÷ snowplows = miles per snowplow

2,256 miles ÷ 24 snowplows = **94 miles**

$$\begin{array}{r} 94 \\ 24\overline{)2,256} \\ 2\ 16 \\ \hline 96 \\ 96 \\ \hline 0 \end{array}$$

Examples 1 and 2. on pages 64 and 65, are really discussing the same situation. In Example 1, the number of plows and miles for each plow are given; you are asked to find the total number of miles that can be covered, and you should multiply. In Example 2, the total number of miles are given, as well as the number of plows to be used. In this case, you are looking for the amount of each piece, the miles for each snowplow, and should divide.

For each problem, write the necessary information into the correct boxes, decide whether to multiply or divide, and then solve the problem.

1. A supermarket sold 78 cartons of Dixie cups. There were 50 cups in every carton. How many cups were sold?

2. Juanita spends an average of $16 a day for food for her family. How much does she spend during the 30-day month of June?

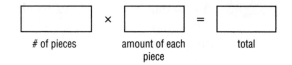

3. The *Washington Post's* morning edition was 140 pages long, and the evening edition was 132 pages long. Of each edition, 780,000 copies were printed. How many pages of newsprint were needed to print the morning edition?

4. Fernando's pickup truck gets 18 miles per gallon. How many miles can he drive on 21 gallons of gasoline?

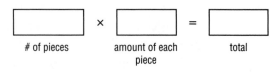

5. In one non-leap year, 46,720 people died in car accidents. What was the average number of deaths each day?

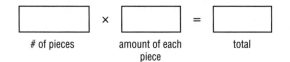

6. A factory produces 68,400 nails a day. Every box is packed with 150 nails. How many boxes does the factory need in one day?

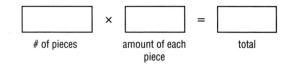

Using Diagrams When Deciding to Multiply or Divide

Drawing a picture or diagram is one important strategy for deciding whether to multiply or divide in order to solve a word problem. Look at how a picture or diagram could have helped you solve Example 1 from page 64.

EXAMPLE 1 Each of the city's 24 snowplows can plow 94 miles of road a day. If all snowplows are running, how many miles of road can be plowed by the city plows in one day?

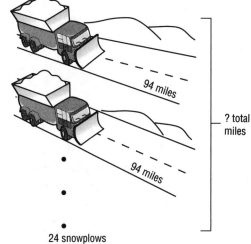

94 miles

94 miles

? total miles

24 snowplows

STEP 1 *question:* How many miles of road can be plowed?

STEP 2 *necessary information:*
24 snowplows, 94 miles of road

STEP 3 Draw a diagram and decide what arithmetic operation to use.

The diagram shows each of the 24 snowplows plowing 94 miles of road. To find the total miles, you should multiply.

STEP 4 Do the arithmetic.

24 × 94 = **2,256 total miles**

STEP 5 Estimate to make sure the answer is sensible.

Over 20 snowplows must each plow nearly 100 miles. An answer near 2,000 miles makes sense.

$$\begin{array}{r} 94 \\ \times\ 24 \\ \hline 376 \\ 188 \\ \hline 2{,}256 \end{array}$$

EXAMPLE 2 A pint of floor wax covers 2,400 square feet of floor. How many pints of floor wax are needed to wax the 168,000-square-foot floor of the airline terminal?

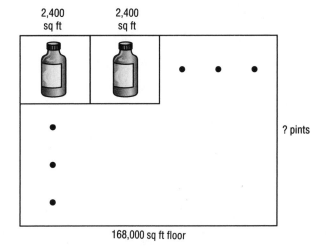

2,400 sq ft

2,400 sq ft

? pints

168,000 sq ft floor

STEP 1 *question:* How many pints of floor wax are needed?

STEP 2 *necessary information:* 1 pint per 2,400 square feet, 168,000 sq ft

STEP 3 Draw a diagram and decide what arithmetic operation to use.

The diagram shows that the 168,000-square-foot airline terminal floor must be divided into sections, each 2,400 square feet (the amount covered by one pint of floor wax). To find the number of pints of floor wax, you should divide.

STEP 4   Do the arithmetic.

168,000 ÷ 2,400 = **70 pints of wax**

$$
\begin{array}{r}
70 \\
2{,}400\overline{)168{,}000} \\
168\,00 \\
\hline
00 \\
00 \\
\hline
\end{array}
$$

STEP 5   Check to be sure the answer is sensible.

You must wax almost 200,000 square feet divided into 2,500 square feet sections. 200,000 ÷ 2,500 = about 80, so 70 pints makes sense.

Each word problem is followed by two diagrams with short explanations. One choice of a diagram and explanation gives you a reason to multiply to find the answer. The other gives you a reason to divide to find the answer. Circle the letter of the correct explanation and solve the problem.

1. Larry feeds each of his 380 laboratory animals 5 ounces of food pellets a day. How many ounces of food pellets does he need for one day?

   a. Since each animal eats 5 ounces, multiply to find the total ounces needed.

   b. To find the amount of food available, divide the total animals (380) by the amount of food available for each.

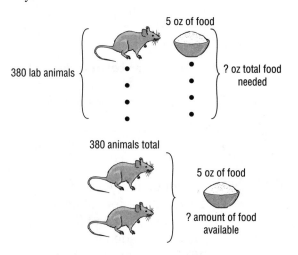

2. Tickets to the play were $60 each. At the end of the night, Janet counted $31,020 in receipts for the performance. How many people bought tickets for the play?

a. To find the number of people, multiply the cost of a ticket ($60) by the receipts ($31,020).

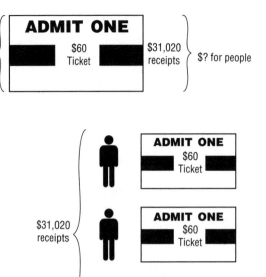

b. To find the number of people, divide the amount of receipts ($31,020) by the price of one ticket ($60).

3. After filling up her gasoline tank, Jean drove 260 miles. After the drive, she refilled her tank with 13 gallons of gasoline. On the average, how many miles was she able to drive on a gallon of gasoline?

a. To find the total miles, multiply the miles for a car (260) by the gallons (13).

260 miles × 13 gallons = ? total miles

b. To find the miles per gallon, divide the miles (260) by the number of gallons (13).

260 miles

13 gallons

? miles

1 gallon

4. Sonya planted 8 rows of tomatoes in her truck garden. If she planted 48 plants in each row, how many tomato plants did she plant?

a. Since there are 8 rows and 48 plants in each row, you should multiply.

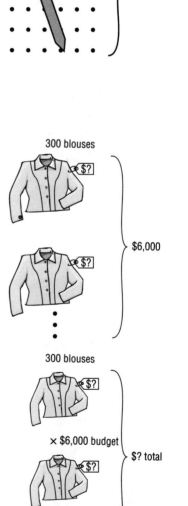

48 plants
48
48
48
48
48
48
48

? tomato plants

b. To find the number of plants, divide 48 plants by 8 rows.

48 plants total

? plants

5. Shirley, a buyer for a major department store, has a budget of $6,000 to buy 300 blouses. What is the most she can pay per blouse?

a. To find the most Shirley can pay for one blouse, divide the total budget ($6,000) by the number of blouses (300).

300 blouses

$?
$?

$6,000

b. To find the total paid, multiply the budget ($6,000) by the number of blouses (300).

300 blouses

$?

× $6,000 budget

$?

$? total

Underline the correct phrase by using the solution given after each word problem. The first one is done for you.

6. A football television contract for $78,000,000 is to be _____ 60 colleges. How much will each college receive?

$$60 \text{ colleges } \overline{)\begin{array}{c} \$\ 1{,}300{,}000 \\ \$78{,}000{,}000 \end{array}}$$ (<u>divided equally among</u>, given to each of)

7. Danika bought 4 skirts for $24 _____. How much did she spend?

$$\begin{array}{r} \$24 \\ \times\ 4 \text{ skirts} \\ \hline \$96 \end{array}$$ (total, each)

8. There were 36,000 trees _____ in the state forest before the fire. During the fire, 48 square miles of forest burned. How many trees were destroyed in all?

$$\begin{array}{r} 36{,}000 \text{ trees} \\ \times\ \quad 48 \text{ square miles} \\ \hline 1{,}728{,}000 \text{ trees} \end{array}$$ (total, per square mile)

9. During an average 12-hour workday, a fast-food restaurant sold 3,852 hamburgers.

$$12 \text{ hours } \overline{)\begin{array}{c} 321 \text{ hamburgers} \\ 3{,}852 \text{ hamburgers} \end{array}}$$ (How many hamburgers were sold in a week?, On average, how many hamburgers were sold per hour?)

10. A cafeteria serves 3,820 _____ a day, with each person being served an 8-fluid ounce portion of soup. How many fluid ounces of soup must be made in one day?

$$\begin{array}{r} 3{,}820 \\ \times\ \quad 8 \text{ fl oz of soup} \\ \hline 30{,}560 \text{ fl oz of soup} \end{array}$$ (fluid ounces of soup, people)

11. The Gourmet Coffee Shop grinds its own coffee beans for its coffee. One pound of coffee beans is used to make 30 cups of coffee. In one week the shop sold 3,180 cups of coffee.

$$30 \text{ cups per pound } \overline{)\begin{array}{c} 106 \text{ pounds} \\ 3{,}180 \text{ cups} \end{array}}$$ (On average, how many cups of coffee were sold each day?, How many pounds of coffee beans were needed to make enough coffee?)

MULTIPLICATION AND DIVISION WORD PROBLEMS: DECIMALS AND FRACTIONS

Solving Decimal Multiplication and Division Word Problems

Fraction and decimal word problems are solved in the same ways as word problems using whole numbers. Estimation can be very helpful with these problems.

EXAMPLE 1   Gasoline costs $3.499 per gallon. How much do 18 gallons of gasoline cost?

STEP 1   *question:* How much do 18 gallons of gasoline cost?

STEP 2   *necessary information:* $3.499 per gallon, 18 gallons

STEP 3   *diagram:*

18 gallons	×	$3.499 per gallon	=	
# of pieces		amount of each piece		total

STEP 4   price of each gallon × number of gallons = total cost

$3.499 × 18 = 62.982 = **$62.98**

(In money problems that have answers containing more than two decimal places, you should round your answer to the nearest cent.)

$$
\begin{array}{r}
3.499 \\
\times\ \ 18 \\
\hline
27992 \\
3499\ \ \\
\hline
62.982
\end{array}
$$

STEP 5   *estimation:* 20 × 3 = 60

This estimation shows that the answer is sensible.

EXAMPLE 2   Gil's car gets 28.6 miles per gallon. Last month he drove 943.8 miles. How many gallons of gasoline did he need for the month?

STEP 1   *question:* How many gallons of gasoline did he need for the month?

STEP 2   *necessary information:* 28.6 miles per gallon, 943.8 miles

STEP 3   *diagram:*

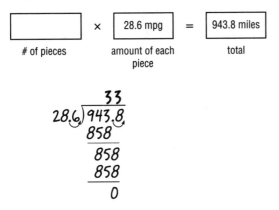

STEP 4   total miles ÷ miles per gallon = gallons

943.8 miles ÷ 28.6 miles per gallon = **33 gallons**

STEP 5   *estimation:* 900 ÷ 30 = 30

> **Calculator:** Your calculator can be very useful when multiplying or dividing decimals. But you must be careful about putting the decimal in the correct place. It is important that you estimate the answer to your problem before doing the calculator work, so you will know that you entered the numbers correctly on the calculator.
>
> If you made a mistake entering the decimal point and got a result of 3.3 gallons, your estimate should alert you.

For each problem, use the diagram to help you decide whether to multiply or divide. Then solve the problem. Remember to write the label of the answer and to round all money problems to the nearest cent.

1. A runner ran an average of 6.5 minutes per mile for a race that had 242 official entrants. If the race was 6.2 miles long, how long did it take him or her to run it?

2. A nonprofit food co-op bought a 40-pound sack of onions for $23.20. How much will the co-op members pay per pound if the onions are sold at cost?

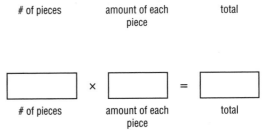

3. Katie filled her gasoline tank and drove 159.75 miles. After the ride, she filled her tank again with 7.1 gallons of gasoline. On the average, how many miles per gallon did she get on the trip?

# of pieces	×	amount of each piece	=	total

4. A beef round roast costs $2.29 per pound. How much is a 4.67-pound roast?

# of pieces	×	amount of each piece	=	total

5. Four roommates share their food bill equally. Last month they spent $372.36 for food and $850.00 for rent. How much did each of them pay for food?

# of pieces	×	amount of each piece	=	total

6. A sales representative, working on commission, earned $118.56 during an 8-hour workday. On the average, how much did the representative earn each hour?

# of pieces	×	amount of each piece	=	total

7. It cost $0.85 per person to ride the city bus. At the end of the day, the driver emptied the cash box and deposited $167.45 in fares. How many passengers rode the bus that day?

# of pieces	×	amount of each piece	=	total

8. Lynelle spent $3.65 for transportation every work day. Last year she worked 239 days. How much did she spend on transportation during work days last year?

# of pieces	×	amount of each piece	=	total

Solving Fraction Multiplication Word Problems

When you multiply two whole numbers, the answer is larger than either number. But when you multiply a number by a fraction smaller than 1, the answer is smaller than the original number. For example, $21 \times \frac{2}{3} = 14$.

Multiplication and division word problems with fractions often seem confusing. When you multiply by a fraction, you may end up with a smaller number, and when you divide by a fraction, you may end up with a larger number. This is the opposite of what you have come to expect with whole numbers.

The following chart should help you remember when to expect a larger or smaller answer when multiplying or dividing.

When Multiplying a Number By:	Your Answer Will Be:	Example:
a number greater than 1	larger than the number	$36 \times 2 = 72$
1	the same as the number	$36 \times 1 = 36$
a fraction less than 1	smaller than the number	$\overset{9}{36} \times \frac{3}{\cancel{4}_1} = 27$
(**Remember:** *An improper fraction is greater than 1. For example,* $36 \times \frac{4}{3} = 48$.)		

When Dividing a Number By:	Your Answer Will Be:	Example:
a number greater than 1	smaller than the number	$36 \div 2 = \overset{18}{36} \times \frac{1}{\cancel{2}_1} = 18$
1	the same as the number	$36 \div 1 = 36$
a fraction less than 1	larger than the number	$36 \div \frac{3}{4} = \overset{12}{36} \times \frac{4}{\cancel{3}_1} = 48$
(**Remember:** *Dividing by an improper fraction is the same as multiplying by a fraction less than 1. For example,* $36 \div \frac{4}{3} = \overset{9}{36} \times \frac{3}{\cancel{4}_1} = 27$.)		

The most common key word in fraction multiplication problems is *of*—as in *finding a fraction of* something. Some people confuse these problems with division because they require you to find a piece of something. The example below illustrates why you multiply when you find a fraction of a quantity.

$$\text{Find } \frac{1}{2} \text{ of } 6.$$

You already know that this is 3. Multiplying by $\frac{1}{2}$ gives the same result as dividing by 2. When you multiply the two numbers, you really multiply the numerators (the numbers above the line) and divide by the denominator (the number below the line).

$$\frac{1}{2} \times 6 = \frac{1}{\underset{1}{2}} \times \frac{\cancel{6}^3}{1} = \frac{3}{1} = 3$$

The following examples show you how to solve multiplication word problems that require you to find a fraction of a quantity.

EXAMPLE 1   Bernie's Service Station inspected 20 cars yesterday. Of the 20 cars inspected, $\frac{1}{5}$ failed the inspection. How many cars failed the inspection?

STEP 1   *question:* How many cars failed the inspection?

STEP 2   *necessary information:* 20 cars, $\frac{1}{5}$ of the cars

STEP 3   *key word:* of

   fraction (of) × total = part

STEP 4 $\frac{1}{5} \times 20$ cars = cars that failed

   $\frac{1}{\cancel{5}_1} \times \frac{\cancel{20}^4}{1} = \mathbf{4\ cars}$

Some multiplication word problems that involve fractions do not have the key word *of.* These problems can be recognized as multiplication, since you are usually given the size of one item and asked to find the size of many. To go from one to many, you should multiply.

EXAMPLE 2   In a high school, class periods are $\frac{3}{4}$ hour long. Luz works as a classroom aide 8 periods a day. How many hours does she work as a classroom aide per day?

STEP 1   *question:* How many hours does she work as a classroom aide per day?

STEP 2   *necessary information:* $\frac{3}{4}$ hour, 8 periods

STEP 3   You are given the length of one class period ($\frac{3}{4}$ hour) and are asked to find the total length of many class periods (8 periods). Therefore, you should multiply.

STEP 4   8 periods × $\frac{3}{4}$ hour = $\frac{^2\cancel{8} \times \frac{3}{\cancel{4}_1}} = \mathbf{6\ hours}$

When you are working with a word problem and have to decide whether to multiply or divide, it is especially helpful to use **Step 5: Check to see that your answer is sensible.**

In Example 2, if you had mistakenly divided 8 by $\frac{3}{4}$, your answer would have been $10\frac{2}{3}$ hours. ($8 \div \frac{3}{4} = 8 \times \frac{4}{3} = 10\frac{2}{3}$) Would it make sense to say that 8 periods, each lasting less than 1 hour, would total $10\frac{2}{3}$ hours?

In the following problems, underline the necessary information. Then solve the problem.

1. In Chicago last year $\frac{2}{3}$ of the precipitation was rain. According to the chart, how many inches of rain fell in Chicago?

Precipitation	
Baltimore	27 inches
Chicago	36 inches
New York	40 inches

2. Of the car accidents in the state last year, $\frac{7}{8}$ were in urban areas. There were 23,352 car accidents in the state last year. How many accidents were in urban areas?

3. How much do 10 of the boxes shown at the right weigh?

$7\frac{2}{3}$ pounds

4. Brad's dog, Cedar, eats $\frac{2}{3}$ can of dog food and two dog biscuits a day. How many cans of dog food will Brad need to feed Cedar for 12 days?

5. A candy bar contains $1\frac{1}{8}$ ounces of peanuts. How many ounces of peanuts are in $3\frac{1}{2}$ candy bars?

6. Only $\frac{1}{2}$ cup of a new concentrated liquid detergent is needed to clean a full load of laundry. How much detergent is needed to clean $\frac{1}{2}$ of a load of laundry?

7. In a $\frac{1}{4}$-pound hamburger, $\frac{2}{5}$ of the hamburger meat was fat. How many pounds of fat was in the hamburger?

8. A space shuttle was traveling 17,000 miles per hour. How far did it travel in $2\frac{1}{2}$ hours?

Solving Fraction Division Word Problems

Remember that the second number in a fraction division problem will be inverted (turned upside down). Therefore, it is very important that the total amount that is being divided is always the first number that you write when solving such a problem. However, even though the total amount comes first when you are solving the fraction division problem, it does not always appear first in a word problem.

EXAMPLE 1　Joyce made dinner for nine people. She divided a $\frac{1}{4}$-pound stick of butter equally among them. How much butter did each person receive?

STEP 1　*question:* How much butter did each person receive?

STEP 2　*necessary information:* $\frac{1}{4}$ pound, 9 people

STEP 3　*key words:* divided equally, each

Nine people are sharing the butter. To find how much butter one person receives, you should divide.

STEP 4　total butter ÷ number of people = butter per person

$\frac{1}{4}$ pound ÷ 9 people $= \frac{1}{4} \times \frac{1}{9} = \mathbf{\frac{1}{36}}$ **pound per person**

Remember: The total amount is not always the largest number.

Many division word problems contain the concept of cutting a total into pieces. If you are given the size of the total, you should divide to find a part—either the number of pieces or the size of each piece.

EXAMPLE 2　Quality Butter Company makes butter in 60-pound batches. Each batch is then cut into $\frac{1}{4}$-pound sticks of butter. How many sticks of butter are made from each batch?

STEP 1　*question:* How many sticks of butter are made from each batch?

STEP 2　*necessary information:* 60-lb batches, $\frac{1}{4}$-lb sticks

STEP 3　*key words:* cuts, each

total amount ÷ size of each piece = number of pieces

STEP 4　60-pound batches ÷ $\frac{1}{4}$-pound sticks $= \frac{60}{1} \div \frac{1}{4}$

$= \frac{60}{1} \times \frac{4}{1}$

$= 60 \times 4$

$= \mathbf{240}$ **sticks**

Underline the necessary information in each problem below. Then solve the problem.

1. A box is $22\frac{1}{2}$ inches deep. How many books can be packed in the box if each book is $\frac{5}{8}$ inch thick?

2. Gloria is serving a dinner for 13 people. She is cooking a $6\frac{1}{2}$-pound roast. How much meat would each person get if she divides the roast equally?

3. A bookstore gift wraps books using $2\frac{1}{4}$ feet of ribbon for each book. How many books can the store gift wrap from a roll of ribbon $265\frac{1}{2}$ feet long?

4. A container contains $8\frac{1}{2}$ pounds of mashed potatoes. Linh, who works in a cafeteria, divides the potatoes into servings the size shown at the right. How many servings can she make from the container of potatoes?

Mashed Potatoes

$\frac{1}{2}$ pound

5. A can of Diet Delight peaches contains $9\frac{3}{4}$ ounces of peaches. If one can is used for three equal servings, how large would each serving be?

6. Tatiana wants to divide the drawing of the garden shown at the right into $1\frac{1}{2}$-foot-wide sections. How many sections can she make?

|← 12 feet →|

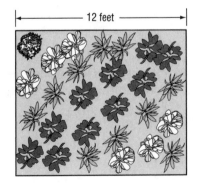

Solving Fraction Multiplication and Division Word Problems

Underline the necessary information and decide whether to multiply or divide. Then solve the problem.

1. A music practice room is used 12 hours a day. If each practice session is $\frac{3}{4}$-hour long, how many sessions are there in a day?

2. At full production, a car rolls off the assembly line every $\frac{2}{3}$ hour. At this rate, how long does it take to produce 30 cars?

3. At full production, a car rolls off the assembly line every $\frac{2}{3}$ hour. At this rate, how many cars are produced in 24 hours?

4. A consumer group claims that $\frac{2}{3}$ of all microwave ovens are defective. The chart shows oven sales in one state last year. According to the consumer group's findings, how many microwave ovens sold in this state would be defective?

Oven Sales	
Microwave	26,148
Regular	59,882

5. On a wilderness hike, six hikers had to share $4\frac{1}{2}$ pounds of chocolate bar and $1\frac{3}{4}$ pounds of dry milk. If it was cut equally, how much chocolate did each hiker receive?

6. A recipe calls for the ingredients at the right. If a cook wants to double the recipe, how much baking soda will he or she need?

1 tsp baking powder

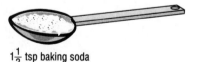

$1\frac{1}{2}$ tsp baking soda

$\frac{1}{2}$ tsp salt

Put a check next to the correct question to complete each problem below.
Use the solution given after each word problem.

7. For her printer, Awilda bought a toner cartridge that contains 3.5 ounces of toner. The toner is supposed to last for 5,000 copies.

   (_____ On the average, how many copies can be made per ounce of toner?)

   (_____ On the average, how much toner is used for each copy?)

$$\begin{array}{r} 0.0007 \text{ oz} \\ 5{,}000\overline{)3.5000} \end{array}$$

8. Elena used to pay $0.13 a minute to call her mother long distance. Since she changed to a new phone card, she now pays $0.09 a minute. This month she talked to her mother for 56 minutes.

   (_____ How much will she spend on the phone calls with her mother?)

   (_____ How much did she save on calls to her mother with her new phone card?)

$$\begin{array}{r} \$0.04 \\ \times\ 56 \\ \hline \$2.24 \end{array}$$

9. In order to cut down the amount of fat in her family's diet, Alejandrina decided to use only $\frac{2}{3}$ of the butter called for in a recipe. Her recipe for brownies originally called for $\frac{1}{2}$ cup of butter.

   (_____ How much butter should she use to double the brownie recipe?)

   (_____ How much butter should she use for her new brownie recipe?)

$\frac{1}{2} \times \frac{2}{3} = \frac{1}{3}$ cup

10. A package of 30 fig bars weighs 16 ounces.

   (_____ What is the total weight of 16 packages of fig bars?)

   (_____ How much does a single fig bar weigh?)

$$\begin{array}{r} 0.53\ \text{oz} \\ 30\overline{)16.000} \\ \underline{150} \\ 100 \\ \underline{90} \\ 100 \end{array}$$

Solving Multiplication and Division Word Problems

For each problem, circle the letter of the correct answer. Round money problems to the nearest cent and other decimal problems to the nearest hundredth.

1. Forty pounds of mayonnaise were packed in jars that weighed $\frac{1}{8}$ pound and could hold $\frac{5}{8}$ pound of mayonnaise. How many jars were needed to pack all the mayonnaise?

    a. 25 jars

    b. 64 jars

    c. 41 jars

    d. 320 jars

    e. 5 jars

2. A 32-square-foot piece of $\frac{1}{4}$-inch-thick Cedar flakeboard costs $20.80. How much does it cost per square foot?

    a. $52.80

    b. $11.20

    c. $0.65

    d. $12.80

    e. $5.20

3. To cover the cost of the prizes, a VFW Post had to sell at least $\frac{1}{6}$ of the raffle tickets. They had 3,000 raffle tickets printed. How many tickets did they have to sell?

    a. 5,000 raffle tickets

    b. 500 raffle tickets

    c. 1,800 raffle tickets

    d. 18,000 raffle tickets

    e. none of the above

4. The trainer of the championship baseball team was voted $\frac{3}{5}$ of a winner's share. If a winner's share is $17,490 and there were 40 shares, how much money did the trainer receive?

    a. $29,155

    b. $10,494

    c. $437.25

    d. $3,498

    e. $728.75

5. To make one apron, Janice needs $\frac{2}{3}$ yard of cloth. She has a roll of cloth $7\frac{1}{3}$ yards long. If she doesn't waste any cloth, how many aprons can she make by cutting and using the entire roll of cloth?

   a. $4\frac{8}{9}$ aprons

   b. 4 aprons

   c. 5 aprons

   d. 11 aprons

   e. 8 aprons

6. Max used 8.1 gallons of gasoline when he drove 263.1 miles in 4.5 hours. What was his average speed for the trip? (Round to the nearest tenth.)

   a. 31.1 miles per hour

   b. 58.3 miles per hour

   c. 58.4 miles per hour

   d. 58.5 miles per hour

   e. 31.2 miles per hour

7. The New Organic Foods Company shipped 1,410 boxes of Date Granola cereal to the Healthy Foods Supermarket. The cereal boxes were packed into cartons for shipping. If 30 boxes of cereal were packed in each carton, how many cartons were needed to ship the boxes of cereal?

   a. 423 cartons

   b. 470 cartons

   c. 47 cartons

   d. 43 cartons

   e. none of the above

8. Ingrid ran in a race from Templeton to Redfield. One kilometer is equal to 0.62 mile. How many miles did she run?

   a. 24 miles

   b. 9.3 miles

   c. 24.19 miles

   d. 4.13 miles

   e. none of the above

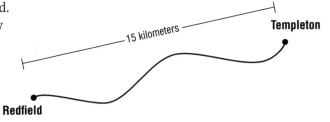

USING PROPORTIONS

What Are Ratios?

A **ratio** is a comparison of two groups. Ratios can be written in a number of ways.

A small luncheonette has 8 chairs for 2 tables. The ratio of chairs to tables can be written three ways.

8 chairs for 2 tables

8 to 2, more commonly written as 8:2

$\dfrac{8\ \text{chairs}}{2\ \text{tables}}$

In the rest of this book, you will use only the third way of writing a ratio, the fraction form.

Note: Always write labels for both the top and the bottom of the ratio.

Write the following relationships as ratios in the fraction form. The first problem has been done for you.

1. 1 customer bought 6 cans of tomato soup. $\dfrac{1\ customer}{6\ cans}$

2. 2 teachers worked with 30 students.

3. Phi Hung earned 120 dollars in 8 hours.

4. Yvette drove 38 miles on 2 gallons of gasoline.

5. The company provided 3 buses for 114 commuters.

What Are Proportions?

A **proportion** expresses two ratios that have the same value. In arithmetic, you have studied these as equivalent fractions. For example, $\frac{75}{100} = \frac{3}{4}$.

EXAMPLE 1 The center of the city has 1 bus stop every 3 blocks. Therefore, over a distance of 6 blocks, there are 2 bus stops.

$$\frac{1 \text{ bus stop}}{3 \text{ blocks}} = \frac{2 \text{ bus stops}}{6 \text{ blocks}}$$

EXAMPLE 2 The ratio of women to men working at the Small Motors Repair Shop is 3 women to 4 men. If there are 8 men working at the repair shop, how many women work there?

One of the numbers of the proportion is not given: the number of women working at the repair shop. Therefore, when the proportion is written, a placeholder is needed in the place where the number of women should be written. The letter *n*, standing for a number, is used as the placeholder, but any letter could be used.

$$\frac{3 \text{ women}}{4 \text{ men}} = \frac{n \text{ women}}{8 \text{ men}}$$

Finding the number that belongs in place of the *n* is called **solving a proportion.** Look at two methods that can be used to solve a proportion.

Method 1: Multiplication

STEP 1 *question:* How many women work there?

STEP 2 *necessary information:* 3 women, 4 men, 8 men

STEP 3 Write a proportion based on the problem.

$$\frac{3 \text{ women}}{4 \text{ men}} = \frac{n \text{ women}}{8 \text{ men}}$$

Notice that both denominators have been filled in and that $4 \times 2 = 8$.

$$\frac{3 \times \square}{4 \times 2} = \frac{n}{8}$$

Since proportions are equivalent fractions, you multiply the numerator and the denominator by the same number. In this problem, the number is 2.

$$\frac{3 \times 2}{4 \times 2} = \frac{6}{8}$$

STEP 4 Therefore, the value of *n* is $3 \times 2 = $ **6 women.**

There are cases when Method 1 does not work as well. This is especially true when a problem contains a decimal or a fraction, or when the numbers are not simple multiples of each other. In these problems, Method 2 is quite useful.

Method 2: Cross Multiplication

STEP 1 *question:* How many women work there?

STEP 2 *necessary information:* 3 women, 4 men, 8 men

STEP 3 Write the proportion.

$$\frac{3 \text{ women}}{4 \text{ men}} = \frac{n \text{ women}}{8 \text{ men}}$$

$$\frac{3}{4} \diagdown\!\!\!\!\diagup \frac{n}{8}$$

STEP 4 Cross multiply. Multiply the numbers that are on a diagonal.

$$4 \times n = 3 \times 8$$
$$4n = 24$$

> **Note:** The letter is usually written on the left side of the equation. Also, $4n$ means the same as 4 times n. It is not necessary to write the multiplication sign, $\times$.

To find n, the number of women, divide the number standing alone by the number next to the letter.

$$n = \frac{24}{4} = 6$$

$n = $ **6 women**

Notice that when you write a proportion, the labels must be consistent. For example, if *women* is the label of the top of one side of a proportion, it must be on the top of the other side.

EXAMPLE 3 Chin has seen 6 movies in the last 9 months. At this rate, how many movies will she see in 12 months?

STEP 1 *question:* How many movies will she see in 12 months?

STEP 2 *necessary information:* 6 movies, 9 months, 12 months

STEP 3 Write the proportion.

$$\frac{12 \text{ months}}{n \text{ movies}} = \frac{9 \text{ months}}{6 \text{ movies}}$$

$$\frac{12}{n} \diagdown\!\!\!\!\diagup \frac{9}{6}$$

STEP 4 Cross multiply. Then divide.

$$9 \times n = 12 \times 6$$
$$9n = 72$$

$n = $ **8 movies**

$$n = \frac{72}{9} = 8$$

EXAMPLE 4 Sandy read that she should cook a roast 20 minutes for each half pound. How large a roast could she cook in 90 minutes?

STEP 1 *question:* How large a roast could she cook in 90 minutes?

STEP 2 *necessary information:* $\frac{1}{2}$ pound, 20 minutes, 90 minutes

STEP 3 Write the proportion.

$$\frac{\frac{1}{2}\text{ pound}}{20\text{ minutes}} = \frac{n\text{ pounds}}{90\text{ minutes}}$$

$$\frac{\frac{1}{2}}{20} \diagdown\!\!\!\!\diagup \frac{n}{90}$$

STEP 4 Cross multiply. Then divide.

$$20 \times n = 90 \times \frac{1}{2}$$

$$n = \textbf{a } 2\frac{1}{4}\textbf{-pound roast}$$

$$20n = 45$$

$$n = \frac{45}{20} = 2\frac{1}{4}$$

Solve the following proportions for *n*.

1. $\dfrac{160\text{ miles}}{5\text{ hours}} = \dfrac{n\text{ miles}}{10\text{ hours}}$

2. $\dfrac{12\text{ cars}}{32\text{ people}} = \dfrac{3\text{ cars}}{n\text{ people}}$

3. $\dfrac{n\text{ dollars}}{8\text{ quarters}} = \dfrac{6\text{ dollars}}{24\text{ quarters}}$

4. $\dfrac{42\text{ pounds}}{n\text{ chickens}} = \dfrac{14\text{ pounds}}{4\text{ chickens}}$

5. $\dfrac{28{,}928\text{ people}}{8\text{ doctors}} = \dfrac{n\text{ people}}{1\text{ doctor}}$

6. $\dfrac{\$24.39}{1\text{ shirt}} = \dfrac{n\text{ dollars}}{6\text{ shirts}}$

7. $\dfrac{\$47.85}{3\text{ shirts}} = \dfrac{n\text{ dollars}}{10\text{ shirts}}$

8. $\dfrac{3\text{ minutes}}{\frac{1}{2}\text{ mile}} = \dfrac{n\text{ minutes}}{5\text{ miles}}$

9. $\dfrac{575\text{ passengers}}{n\text{ days}} = \dfrac{1{,}725\text{ passengers}}{21\text{ days}}$

10. $\dfrac{7\text{ blinks}}{\frac{1}{10}\text{ minute}} = \dfrac{n\text{ blinks}}{10\text{ minutes}}$

Using Proportions to Solve Word Problems

Proportions can be used when you are unsure of whether to multiply or divide.

The following examples show how to write proportions to solve multiplication and division word problems.

EXAMPLE 1 There are 16 cups in a gallon. At the picnic, Carmella poured 5 gallons of cola into paper cups that each held 1 cup of soda. How many cups did she fill?

STEP 1 *question:* How many cups did she fill?

STEP 2 *necessary information:* 16 cups in a gallon, 5 gallons, 1 cup

STEP 3 Write the proportion.

labels for proportion: $\dfrac{\text{cups}}{\text{gallons}}$

$$\frac{16 \text{ cups}}{1 \text{ gallon}} = \frac{n \text{ cups}}{5 \text{ gallons}}$$

$$\frac{16}{1} \diagup\!\!\!\!\diagdown \frac{n}{5}$$

STEP 4 Cross multiply.

(*n* means the same as 1 × *n*. From now on you don't need to write the 1 and the × sign, so you write $n = 16 \times 5$.)

$$n = 16 \times 5$$

$$n = 80$$

$n = \textbf{80 cups}$

Remember: When you write the labels for a proportion, it doesn't matter which category goes on top. But once you make a decision, you must stick with it. Once you put *cups* on the top of one ratio, you must keep *cups* on top of the other.

---

16 cups in a gallon means the same as $\frac{16 \text{ cups}}{1 \text{ gallon}}$. The 1 will often not appear in these word problems. When writing a proportion, you must determine when a 1 is needed and where it goes. You can do this by first identifying the two labels and then putting numbers in the proportion.

There are a number of word phrases that require that a 1 be used in a ratio.

Phrases	Meaning
27 miles per gallon	$\dfrac{27 \text{ miles}}{1 \text{ gallon}}$
$8 an hour	$\dfrac{\$8}{1 \text{ hour}}$
3 meals a day	$\dfrac{3 \text{ meals}}{1 \text{ day}}$
30 miles each day	$\dfrac{30 \text{ miles}}{1 \text{ day}}$

EXAMPLE 2 At the Boardwalk Arcade, owner Manuel Santos collects 1,380 quarters every day. There are 4 quarters in a dollar. How many dollars does he collect every day?

STEP 1 *question:* How many dollars does he collect every day?

STEP 2 *necessary information:* 1,380 quarters, 4 quarters in a dollar

STEP 3 *labels for proportion:* $\dfrac{\text{dollars}}{\text{quarters}}$

$$\frac{n \text{ dollars}}{1,380 \text{ quarters}} = \frac{1 \text{ dollar}}{4 \text{ quarters}}$$

$$\frac{n}{1,380} \diagup\!\!\!\!\diagup \frac{1}{4}$$

STEP 4 Cross multiply. Then divide.

$$4 \times n = 1,380 \times 1$$

$$n = \textbf{345 dollars}$$

$$4n = 1,380$$

$$n = \frac{1,380}{4} = 345$$

Underline the necessary information. Write proportions for the problems below and solve them.

1. A shipment of vaccine can protect 7,800 people. How many shipments of vaccine are needed to protect 140,400 people living in the Portland area?

2. It costs $340 an hour to run the 1,000-watt power generator. How much does it cost to run the generator for 24 hours?

3. How many ounces of soup (like the can shown at the right) are in a carton containing 28 cans?

4. Jim types 52 words per minute. How many words can he type in 26 minutes?

5. An elementary school nurse used 3,960 Band-Aids last year. There were 180 school days. On the average, how many Band-Aids did he use a day?

6. The company health clinic gave out 5,460 aspirins and 720 antacid tablets last year. How many bottles of aspirin did the clinic use last year if there were 260 aspirins in a bottle?

7. A coal mine produced 126 tons of slag in a week. Trucks removed the slag in 3-ton loads. How many loads were needed to remove all the slag?

8. Cloth is sold by the yard. Carmen bought the piece of cloth shown at the right to make dresses. There are 3 feet in a yard. How many yards of cloth did she buy?

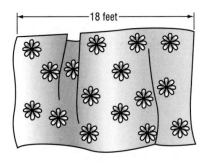

9. A 6-ounce can of water chestnuts contains 26 water chestnuts. Hong uses 3 cans of water chestnuts. How many water chestnuts does she use?

Using Proportions to Solve Decimal Word Problems

You can use proportions to solve decimal multiplication and division word problems. The problems should be set up as if the numbers were whole numbers. Multiply or divide as if you were using whole numbers. Then use the rules for decimal multiplication and division to place the decimal point in the right place. Finally, round the answer if necessary.

EXAMPLE 1   Ray has $45.00 to spend on gasoline. How many gallons can he buy if 1 gallon costs $3.60?

STEP 1   *question:* How many gallons can he buy?

STEP 2   *necessary information:* $45.00, $3.60 for a gallon

STEP 3   *labels for proportion:* $\dfrac{\$}{\text{gallons}}$

$$\frac{\$45}{n \text{ gallons}} = \frac{\$3.60}{1 \text{ gallon}}$$

STEP 4   Cross multiply. Then divide.

$n =$ **12.5 gallons**

$$3.60 \times n = 45 \times 1$$
$$3.60n = 45$$
$$n = \frac{45}{3.60} = 12.5$$

EXAMPLE 2   There are 236.5 milliliters in a cup. A recipe calls for 3 cups of flour. Maria has only metric spoons and measuring cups. How many milliliters of flour does she need for the recipe?

STEP 1   *question:* How many milliliters of flour does she need for the recipe?

STEP 2   *necessary information:* 236.5 milliliters in a cup, 3 cups

STEP 3   *labels for proportion:* $\dfrac{\text{milliliters}}{\text{cups}}$

$$\frac{236.5 \text{ milliliters}}{1 \text{ cup}} = \frac{n \text{ milliliters}}{3 \text{ cups}}$$

STEP 4   Cross multiply.

$n =$ **709.5 milliliters**

$$\frac{236.5}{1} \times \frac{n}{3}$$
$$n = 3 \times 236.5$$
$$n = 709.5$$

Underline the necessary information. Write the labels for the proportion, fill in the numbers, and solve the proportion. Round your answers to the nearest hundredth.

1. There are 25.4 millimeters in an inch. How many inches long is a 100-millimeter cigarette?

2. A kilogram weight is shown at the right. The police seized 36 kilograms of illegal drugs. How many pounds did the drugs weigh?

1 KILOGRAM
2.2 POUNDS

3. Alba worked 35.5 hours last week. She earned $12.62 an hour. How much money did she earn last week?

4. There are 1.09 yards in a meter. Gary runs in an 880-yard race. How many meters does he run?

5. Cindy spent $38.00 on gasoline. The gasoline cost $4.20 per gallon. How many gallons of gasoline did she buy?

6. There are approximately 1.61 kilometers in a mile. The speedometer on Iris's imported car is in kilometers per hour. She does not want to speed. What is 55 miles per hour in kilometers per hour?

7. A Tiger Milk nutrition bar weighs 35.4 grams. The factory processes 9,486 bars in one run. How many grams of Tiger Milk bars are processed?

Using Proportions to Solve Fraction Word Problems

Proportions can be used to solve multiplication and division word problems containing fractions. Though they look complicated when you set them up, they are manageable after you cross multiply.

EXAMPLE 1 A pickup truck can carry a load of $2\frac{3}{4}$ tons of gravel. In one day, the truck delivers 8 loads of gravel to an office building construction site. How many tons of gravel does it deliver to the construction site that day?

STEP 1 *question:* How many tons of gravel does it deliver to the construction site that day?

STEP 2 *necessary information:* a load of $2\frac{3}{4}$ tons, 8 loads

STEP 3 *labels for proportion:* $\dfrac{\text{tons}}{\text{loads}}$

$$\frac{2\frac{3}{4}\text{ tons}}{1\text{ load}} = \frac{n\text{ tons}}{8\text{ loads}}$$

STEP 4 Cross multiply.

$n = \textbf{22 tons}$

$$\frac{2\frac{3}{4}}{1} \diagup\!\!\!\!\diagdown \frac{n}{8}$$

$$n = 8 \times 2\frac{3}{4}$$

$$n = \frac{\overset{2}{\cancel{8}}}{1} \times \frac{11}{\underset{1}{\cancel{4}}} = 22$$

Underline the necessary information in each problem below. Write the proportions and solve the problems.

1. A slicing machine can cut roast beef slices $\frac{1}{16}$ inch thick. The giant sandwich was advertised to contain roast beef slices piled 2 inches thick. How many slices of roast beef were on the sandwich?

2. The La Ronga Bakery baked 1,460 loaves of bread in 1 day. If each loaf contained $1\frac{3}{4}$ teaspoons of salt, how much salt did the bakery use?

3. How many books, each $\frac{7}{8}$ inch thick, can be packed in a box that is 35 inches deep?

4. A can of pears weighs $9\frac{2}{3}$ ounces. There are 16 cans of pears in a carton. How many ounces does a carton of pears weigh?

5. There are 8 cups of detergent in a bottle of Easy Clean detergent. For one load of laundry, $\frac{1}{4}$ cup is all that is needed. How many loads of laundry can be cleaned with a bottle of Easy Clean?

Solving Conversion Word Problems

Have you ever seen this kind of problem?

EXAMPLE 1 Caren has a 204-**inch** roll of masking tape. How many **feet** of molding can she cover with the roll?

This problem is an example of a type of multiplication or division word problem that contains only one number and requires outside information in order to be solved. These are word problems involving **conversions** from one type of measurement to another.

Here is the solution and explanation of the example.

STEP 1 *question:* How many feet?

STEP 2 *necessary information:* 204 inches

Notice that the question asks for a solution that has a different label than what is given in the problem. To solve this, you must know how to convert inches to feet. Then you can set up a proportion to solve the problem. You may need to refer to the conversion chart on page 221.

STEP 3 12 inches = 1 foot

labels for proportion: $\dfrac{\text{inches}}{\text{feet}}$

The conversion will be one side of the proportion.

$$\frac{12 \text{ inches}}{1 \text{ foot}} = \frac{204 \text{ inches}}{n \text{ feet}}$$

$$\frac{12}{1} \diagup\kern-1em\diagdown \frac{204}{n}$$

STEP 4 Cross multiply. Then divide.

$n = \textbf{17 feet}$

$$12 \times n = 204 \times 1$$
$$12n = 204$$
$$n = \frac{204}{12} = 17$$

A diagram can often help you picture a conversion word problem.

EXAMPLE 2 The Spring Lake Day-Care Center gives each of its 12 children a cup of milk for lunch every day. How many quarts of milk does the center use each day?

STEP 1 *question:* How many quarts of milk does the center use each day?

STEP 2 *necessary information:* 12 cups, 1 cup
conversion formula: 4 cups = 1 quart

STEP 3 Decide what arithmetic operation to use.

The diagram shows that you should divide.

number of cups ÷ cups in a quart = number of quarts

STEP 4 Do the arithmetic.

$$12 \text{ cups} \div \frac{4 \text{ cups}}{1 \text{ quart}} = \textbf{3 quarts}$$

STEP 5 Make sure the answer is sensible. Look at the diagram to see that the answer of 3 quarts makes sense.

Use the conversion chart on page 221 to help you write the conversion and the proportion for each problem. Then solve the problem.

1. How many years old is Gloria's 30-month-old daughter?

2. A 200-gallon batch of ketchup was bottled in quart bottles. How many bottles were filled?

3. How many kilometers long is a 10,000-meter road race?

4. Paul's truck can carry a $\frac{1}{2}$-ton load. How many pounds of gravel can it carry?

5. José brought the cream shown at the right to the company picnic. How many ounces of cream did he bring?

6. Mt. Everest is 29,028 feet above sea level. How many miles is that equal to? (Round to the nearest tenth.)

Using Proportions with Fractions and Decimals

Sometimes you will find a multiplication or division word problem in which one number is a decimal and the other is a fraction. The proportion method is very useful in solving this type of word problem.

EXAMPLE   A $\frac{3}{4}$-pound steak cost \$6.75. How much did it cost per pound?

STEP 1   *question:* How much did it cost per pound?

STEP 2   *necessary information:* $\frac{3}{4}$ pound, \$6.75

STEP 3   *labels for proportion:* pound, \$

$$\frac{\frac{3}{4} \text{ pound}}{\$6.75} = \frac{1 \text{ pound}}{\$n}$$

STEP 4   Cross multiply. Then multiply both sides by $\frac{4}{3}$.

$$n = \textbf{\$9.00}$$

$$\frac{\frac{3}{4}}{6.75} \diagup\!\!\!\!\diagdown \frac{1}{n}$$

$$\frac{3}{4}n = 6.75$$

$$n = \overset{2.25}{\cancel{6.75}} \times \frac{4}{\cancel{3}_1}$$

$$n = 9.00$$

If you had not been able to cancel in the example, you would have multiplied the numerators and divided by the product of the denominators. Keep the decimal point in the correct place.

Write the proportions and solve the problems below. Round money problems to the nearest cent.

1. Cloth was being sold at \$12.60 a yard. Lei bought $3\frac{1}{3}$ yards of cloth. How much did she spend?

2. George was told that 13.5 pounds of time-release fertilizer should last $4\frac{1}{2}$ years. How much fertilizer is used each year?

3. Carlo bought $2\frac{2}{3}$ pounds of grapes for \$3.25. How much did the grapes cost per pound?

4. Quality Landscaping charges \$35.65 per cubic yard of screened loam, including delivery and spreading. Scott and Tanya Greenblatt ordered $7\frac{1}{2}$ cubic yards of loam to be delivered to their new home. How much did Quality Landscaping charge the Greenblatts for the loam?

Using Proportions with Multiplication and Division

Underline the necessary information. Write the proportions and solve the problems.

1. A butcher can cut up a chicken in $\frac{1}{12}$ of an hour. How many chickens can be cut up in an 8-hour work day?

2. A nurse can take 8 blood samples in 60 minutes. How long does it take her to take one blood sample?

3. A mile is about 1.6 kilometers. About how many kilometers is a 26-mile marathon?

4. An oil-drilling rig can drill 6 feet in an hour. How far can it drill in 24 hours?

5. A gram is about 0.04 ounce. About how many grams are in a 12-ounce can of pineapple slices?

6. Rosa uses $3\frac{1}{4}$ pounds of pumpkin to make 2 pies. For a fall bake sale, she made 10 pumpkin pies. How many pounds of pumpkin did she use?

7. When the floodgates were opened, 68,000 gallons of water flowed over the dam per hour. How many gallons flowed over the dam in a day?

8. Lace Heaven charges $0.40 per yard for its $\frac{1}{2}$-inch powder pink stretch lace. How much did Zelda spend on $4\frac{1}{4}$ yards of trimming that she bought at Lace Heaven?

9. Super Glue sets in $3\frac{1}{2}$ minutes. In how many seconds does Super Glue set?

10. A 942-page book contained 302,382 words. On the average, how many words were on each page?

11. Top Burger makes a $\frac{1}{4}$-pound hamburger. How many of these hamburgers can be made from 50 pounds of ground beef?

12. Of all the gallon-size milk containers sold in a store, $\frac{2}{5}$ were low-fat milk. The store sold 380 gallons of milk. How many gallons of low-fat milk were sold?

Mixed Word Problems with Whole Numbers

So far, you have worked with word problems that have been divided into two major categories—addition/subtraction problems and multiplication/division problems.

In most situations, you will be faced with the four types of problems mixed together. Always read each problem carefully to get an understanding of the situation it describes. This will help you choose the right arithmetic operation.

Keep these general guidelines in mind:

- when combining amounts   → add

- when finding the difference
  between two amounts   → subtract

- when given one unit of something
  and asked to find several   → multiply

- when asked to find a fraction
  of a quantity   → multiply

- when given the amount for
  several and asked for one   → divide

- when dividing, cutting, sharing   → divide

Working through the following exercises will help sharpen your skills with word problems when the different types are mixed together.

Write the arithmetic operation (addition, subtraction, multiplication, or division) that you would use to solve each problem and the reason for your choice. DO NOT SOLVE!

1. Doreen needs 39 credits to complete her bachelor's degree at the state university. The university charges $265 per credit for tuition. How much will Doreen have to pay in tuition to complete her degree?

2. Alan wrote a 74,200-word manuscript for his new book. His manuscript was 280 pages long. On average, how many words per page were in the manuscript?

3. After laying off 27 workers, Paul still had 168 workers at the hospital. How many workers were there at the hospital before the changes?

4. A department store bought belts for $14 each and sold them for $24. How much profit did it make on each belt?

5. Tomas needs 19 feet of molding for each doorway in his home. The home has 9 doorways. How much molding does Tomas need?

6. Part of Saba Farm's harvest log is shown at the right. How many more ears of corn did they harvest on Thursday than on Wednesday?

	WED	THURS
Corn	476	548
Zucchini	94	129

7. As coach of her soccer team, Althea decided that all 22 players would get equal playing time. With a total of 990 minutes to distribute, how many playing minutes did Althea give to each of her players?

8. At her day-care center, Jana used an entire gallon of juice at snack time for 24 children. On the average, how many fluid ounces of juice did each child receive?

9. The governor's goal is to reduce the state's imports of foreign oil by 25,000 barrels to 90,000 barrels a month. How much oil is the state currently importing a month?

Using Labels to Solve Word Problems

Every number in a word problem has a **label.** Every number *refers* to something. It makes no sense to say simply "7" or "$38\frac{1}{2}$." We need to know—7 *what?* $38\frac{1}{2}$ *what?* Dogs? Miles per hour? Years old? What do these numbers refer to?

Paying careful attention to labels will help you decide whether to add, subtract, multiply, or divide. Look at the following example.

EXAMPLE 1 On Saturday night, Bruce spent $56.50 on dinner and $48.00 for tickets to a play. How much did he spend altogether?

Notice that the labels of both pieces of the necessary information are *dollars*. Also, you can tell that the label of the answer will be in *dollars*. You probably already have figured out that you need to add $56.50 and $48.00 to solve the problem.

Look at the next example.

EXAMPLE 2 In the last election, 35,102 women and 29,952 men voted. How many people voted?

The labels of the necessary information and the answer are different. Does this mean you should not add or subtract?

Whenever the labels of items in a word problem are different, first ask yourself if the different labels can be part of a *broader category* or if they can be *converted* to a common unit (such as from *pounds* to *ounces* or from *years* to *months*). For example, *men* and *women* can both be considered part of the broader category of *people*. Therefore, when all the labels in the problem are the same you can add or subtract to find the answer. For the problem above, you should add 35,102 and 29,952 to get the answer.

> You will often find this pattern in word problems:
>
> When the labels of all the necessary information and the answer are the same, you usually need to *add* or *subtract* to solve.

Using labels to decide whether to multiply or to divide is a bit trickier. However, if you are willing to play a bit with the labels in a problem, you can often make this decision before you have to work with actual numbers. Read the next example.

EXAMPLE 3   Matteo drove 385 miles in 7 hours. How many miles per hour did he average on this trip?

The labels of one piece of the necessary information is *miles*; the label of the other piece is *hours*. The label of the answer is *miles per hour*. Already you may be guessing that you should not add or subtract, for the labels are *not* the same and they can't be converted to a common unit.

The answer will be in *miles per hour*, which can also be written as $\frac{miles}{hour}$ (miles *divided by* 1 hour). Set up a statement using the labels from the problem:

$$\text{miles} \;\square\; \text{hours} = \frac{\text{miles}}{\text{hour}}$$

> **Remember:** *miles per hour* means the ratio $\frac{\text{miles}}{\text{hour}}$

How would you fill in the box? Ask yourself, "What do I have to do to *miles* and *hours* in order to get $\frac{miles}{hour}$?"

You need to divide.

$$385 \text{ miles } \boxed{\div} \; 7 \text{ hours} = \mathbf{55 \frac{\text{miles}}{\text{hour}}}$$

OR

$$385 \text{ miles} \div 7 \text{ hours} = \textbf{55 miles per hour}$$

Now look at the next example.

EXAMPLE 4   Matteo took a 7-hour trip. He averaged 55 miles per hour on the trip. How many miles in all did he drive?

The label of one piece of the necessary information is *hours*; the label of the other piece is *miles per hour*. The label of the answer will be *miles*. Ask yourself, "What would I do to *hours* and *miles per hour* to get *miles*?"

$$\text{hours} \;\square\; \frac{\text{miles}}{\text{hour}} = \text{miles}$$

Try multiplying. Just as with numerical multiplication, you can cancel out common factors (in this case, the label *hour*).

canceling
using words

$$\cancel{\text{hours}} \;\boxed{\times}\; \frac{\text{miles}}{\cancel{\text{hour}}} = \text{miles}$$

canceling
using numbers

$$\cancel{5} \times \frac{3}{\cancel{5}} = 3$$

Because canceling labels leaves you the label to your answer, you know that you should multiply to get the correct answer.

$$\text{hours} \; \boxed{} \; \frac{\text{miles}}{\text{hour}} = \text{miles}$$

$$7 \; \text{hours} \; \boxed{\times} \; \frac{55 \text{ miles}}{1 \text{ hour}} = ? \text{ miles}$$

$$7 \times 55 \text{ miles} = \textbf{385 miles}$$

Now, look at a problem in which you can't cancel the labels.

EXAMPLE 5 Matteo took a 385-mile trip and drove 55 miles per hour. How many hours did he drive?

Look at the expression. $\quad \text{miles} \; \boxed{} \; \frac{\text{miles}}{\text{hour}} = \text{hours}$

Can you convert the labels to a common unit? No. Can you cancel the labels *miles* and *hours*? No. You have one more case you should try.

Remember the procedure for dividing fractions is to invert (flip over) the fraction you are dividing by and then multiply the result. You can do the same thing with labels.

$$385 \text{ miles} \; \boxed{\div} \; \frac{55 \text{ miles}}{1 \text{ hour}} = ? \text{ hours}$$

$$385 \; \text{miles} \times \frac{1 \text{ hour}}{55 \text{ miles}} = ? \text{ hours}$$

$$385 \times \tfrac{1}{55} \text{ hours} = 7 \text{ hours}$$

When you actually do the math, it is easier just to divide, but it is important to know that you can still use the labels to decide which operation to use.

Read each of the following problems. Look at the labels and decide if they can be renamed to a broader category. Then write in the correct operation to solve the problem. Finally, solve the problem.

1. Sam cut an 8-ounce slice from a 20-pound round of cheese. How many ounces of cheese were left?

   necessary information labels: _____ _____

   answer label: _____

   Are the labels different? ____*yes*____

   If so, what is the new label? ___*ounces*___

   _____ $\boxed{}$ _____ = _____
      label   label   label

   Answer: _____

2. Adrienne needed to cut 2 feet from a 72-inch piece of molding. After the cut, how much molding was left?

necessary information labels: _____ _____

answer label: _____

Are the labels different? _____

If so, what is the new label? _____

_____ $\square$ _____ = _____
label   label   label

Answer: _____

3. Maura's hair was $9\frac{1}{2}$ inches long. How long was her hair before she had it cut by $1\frac{3}{4}$ inches?

necessary information labels: _____ _____

answer label: _____

Are the labels different? _____

If so, what is the new label? _____

_____ $\square$ _____ = _____
label   label   label

Answer: _____

Read the following problems. First look at the labels in the problem and see if they can be canceled. If the labels can't be canceled, flip over the second label, and then check if the labels can be canceled. Then write the correct operation in the box. Finally, solve the problem.

4. A ream of paper contains 500 sheets. A box contains 10 reams of paper. How many sheets of paper are in the box?

$\dfrac{\text{sheets}}{\text{ream}}$ $\square$ reams = sheets   Answer: _____

5. It cost $9 to go to the movies. A movie theater collected $5,220 in ticket sales. How many tickets were sold?

dollars $\square$ $\dfrac{\text{dollars}}{\text{ticket}}$ = tickets   Answer: _____

6. An average tomato plant in John's garden yields 8 pounds of tomatoes. How many pounds can he expect from his 14 plants?

plants $\square$ $\dfrac{\text{pounds}}{\text{plant}}$ = pounds   Answer: _____

Not Enough Information

Now that you know how to decide whether to add, subtract, multiply, or divide to solve a word problem, you should be able to recognize a word problem that cannot be solved because not enough information is given.

Look at the following example.

EXAMPLE 1   At her waitress job, Sheila earns $8.50 an hour plus tips. Last week she earned $265.40 in tips. How much did she earn in all last week?

STEP 1   *question:* How much did she earn in all last week?

STEP 2   *necessary information:* $8.50 an hour, $265.40

STEP 3   Decide what arithmetic operation to use.

tips + (pay per hour × hours worked) = total earned

missing information: hours worked

At first glance, you might think that you have enough information since there are two numbers. But when the solution is set up, you can see that you need to know the number of hours Sheila worked to find out what she earned in all.

For each word problem, circle the letter of the information needed to solve the problem.

1. A supermarket sold 350 pounds of bananas at $0.69 a pound. How many pounds of bananas did it have left?

   a. You need to know how much the supermarket paid for the bananas.

   b. You need to know how many pounds of bananas the supermarket started with.

   c. You need to know how much money the supermarket made for each pound of bananas sold.

   d. You have enough information to solve the problem.

2. In one day last year, 2,417 people were born or moved into the state and 1,620 people died or left the state. What was the state's change of population for the day?

   a. You need to know the total population of the state.

   b. You need to know the name of the state.

   c. You need to know exactly how many people were born and exactly how many died.

   d. You have enough information to solve the problem.

3. Roast beef that normally cost $4.59 a pound was marked down $0.60. If Gina paid for a roast with a $20.00 bill, how much change did she receive?

   a. You need to know the weight of Gina's roast beef.

   b. You need to know the total amount of money Gina had.

   c. You need to know how much the supermarket paid for the roast beef.

   d. You have enough information to solve the problem.

4. A loaded truck carrying boxes of books weighed 7,105 pounds at the weigh station. If each box of books weighed 42 pounds, how much did the unloaded truck weigh?

   a. You need to know how many books were in the truck.

   b. You need to know the weight of a single book.

   c. You need to know how many boxes were in the truck.

   d. You have enough information to solve the problem.

5. A bag of 40 snack bars weighs 12 ounces. How much does each snack bar weigh?

   a. You need to know the price of one snack bar.

   b. You need to know how many ounces are in a pound.

   c. You need to know the total price of the entire bag.

   d. You have enough information to solve the problem.

In the following problems, decide whether to add, subtract, multiply, or divide. Then solve the problem. Circle the letter of the correct answer.

6. Mr. Gomez's obituary appeared in a 1998 newspaper. It said that he was 86 years old when he died and had been married for 51 years. In what year was he born?

   a. 1903

   b. 1947

   c. 1861

   d. 1887

   e. 1912

7. An electrician has a piece of cable the length shown at the right. How long a cable would he have if he laid 7 of these pieces end to end?

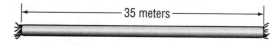

   a. 5 meters

   b. 28 meters

   c. 42 meters

   d. 245 meters

   e. 490 meters

8. The population of San Jose rose by 128,140 people between 2000 and 2010. The population had been 894,943. What was the new population?

   a. approximately 7 times

   b. approximately 8 times

   c. 766,803 people

   d. 1,023,083 people

   e. not enough information given

9. A Christmas light uses 2 watts of electricity. How many lights can be strung on a circuit that can handle a load of 300 watts?

   a. 150 lights

   b. 600 lights

   c. 298 lights

   d. 302 lights

   e. none of the above

10. A telephone cable can handle 12,500 calls at any one time. How many cables are needed to handle a peak load of 87,500 calls?

    a. 100,000 cables

    b. 75,000 cables

    c. 7 cables

    d. 70 cables

    e. none of the above

11. Ravi bought $360 worth of sports equipment and $18 worth of office supplies for the boys' club. Since the boys' club is tax-exempt, he didn't have to pay the sales tax. If he had paid tax, how much would he have spent?

    a. $20

    b. $378

    c. $342

    d. $360

    e. not enough information given

12. A clothing factory produced 8,760 yards of cloth. What was the average production from each of the 60 looms in the factory?

    a. 146 yards

    b. 1,460 yards

    c. 8,700 yards

    d. 8,820 yards

    e. 525,600 yards

13. At Priced to Sell Motors, Len's goal was to sell 20 SUVs a month. Part of Len's sales log is shown at the right. By how much did he exceed his goal in September?

    a. 2 SUVs

    b. 13 SUVs

    c. 17 SUVs

    d. 57 SUVs

    e. 740 SUVs

SUV Sales	
August	19
September	37
October	30
November	21

In the following problems, write a question that matches the solution.

14. In 2008, Maryland at $70,545 was the state with the highest median household income. At $37,790, Mississippi had the lowest median household income in 2008.

$$\$70{,}545 - \$37{,}790 = n$$

15. During the pre-Christmas sale, the price of the new car was slashed from $15,364 to $11,994.

$$\$15{,}364 - \$11{,}994 = n$$

16. In order to be stained, concrete needs to be treated with 4 parts muriactic acid mixed with 1 part water. Jaresh has a 1 pint container of muriactic acid.

$$\frac{4 \text{ parts muriactic acid}}{1 \text{ part water}} = \frac{1 \text{ pint muriactic acid}}{n \text{ pints water}}$$

17. When building a roof, Rosalia knew that she needed 1 square foot of ventilation for every 300 square feet of attic floor space. Her attic was 1,200 square feet.

$$\frac{1 \text{ sq ft ventilation}}{300 \text{ sq ft attic}} = \frac{n \text{ sq ft ventilation}}{1{,}200 \text{ sq ft attic}}$$

18. A base coat of stucco uses $2\frac{1}{2}$ parts clean common sand to 1 part Type N cement. Cesar has a 50-pound bucket of Type N cement.

$$\frac{2\frac{1}{2} \text{ parts sand}}{1 \text{ part Type N cement}} = \frac{n \text{ lb sand}}{50 \text{ lb Type N cement}}$$

19. A drill press can drill a hole accurately to 0.002 inch. The press is set to drill a 0.235-inch hole.

$$0.235 \text{ inch} + 0.002 \text{ inch} = n \text{ inch}$$

20. Abad and Socorro stayed overnight at Motel 5. The room charge was $59.95. The motel taxes were $16.25.

$$\$59.95 + \$16.25 = n$$

Mixed Word Problems: Whole Numbers, Decimals, and Fractions

In the following problems, circle the letter of the correct answer. Round decimals to the nearest hundredth.

1. Sandor bought a roast beef sandwich for $3.78, which included $0.18 tax. What was the cost of the sandwich alone?

   a. $3.96

   b. $3.78

   c. $3.60

   d. $2.10

   e. $0.68

2. In 2010 the estimated population of the United States was 309,187,390. The 2010 estimated population of Russia was 141,927,297. How much greater was the population of the United States than the population of Russia?

   a. 451,115,687 people

   b. 451,112,677 people

   c. 167,260,093 people

   d. 167,260,103 people

   e. none of the above

3. To tie her tomato plants, Emmy cut the string shown at the right into $\frac{3}{4}$-foot long pieces. How many pieces of string did she have to tie her tomatoes?

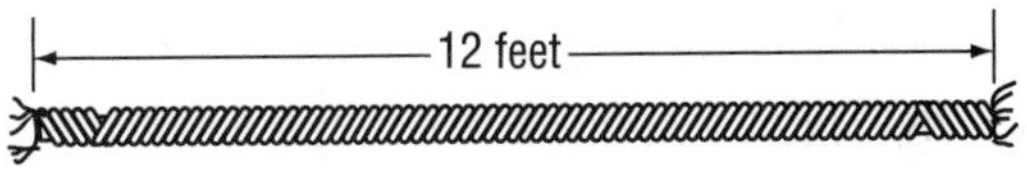

   a. $12\frac{3}{4}$ feet

   b. $11\frac{1}{4}$ feet

   c. 9 pieces

   d. 16 pieces

   e. none of the above

4. Roasted chickens cost $1.95 a pound at the Neighborhood Supermarket. Cali paid $8.70 for a roasted chicken at Neighborhood Supermarket. How much did the chicken weigh? (Round to nearest hundredth.)

   a. 6.75 pounds

   b. 16.97 pounds

   c. 10.65 pounds

   d. 4.46 pounds

   e. 2.24 pounds

5. Jean has a small dressmaking business. She received an order to make matching mother-daughter skirts. She needed $2\frac{1}{2}$ yards of 60-inch wide material for the mother's dress and $\frac{7}{8}$ yard of material for the daughter's dress. How many yards of material did she need to buy?

    a. $3\frac{3}{8}$ yards

    b. $2\frac{6}{7}$ yards

    c. $2\frac{3}{16}$ yards

    d. $1\frac{5}{8}$ yards

    e. none of the above

6. What is the total weight of the two packages of fruit shown at the right?

    a. 1.81 pounds

    b. 1.63 pounds

    c. 2.62 pounds

    d. 1.55 pounds

    e. 0.82 pound

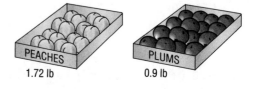

PEACHES 1.72 lb   PLUMS 0.9 lb

7. A serving of Kellogg's Raisin Bran contains 0.26 gram of potassium. How many grams of potassium are in an 11-serving package of Raisin Bran?

    a. 11.26 grams

    b. 10.74 grams

    c. 42.31 grams

    d. 2.86 grams

    e. 0.02 gram

8. The auto repair shop charged Muriel $1,125 to repair her car. She had a $250-deductible insurance policy. How much did the insurance company pay for the repair of her car?

    a. $1,375.00

    b. $875.00

    c. $4.50

    d. $281.25

    e. none of the above

9. Nataliya spent $\frac{1}{3}$ of her paycheck on food and $\frac{1}{4}$ for clothes. Her paycheck was for $414. How much did she spend for food?

a. $138.00

b. $34.50

c. $241.50

d. $1,242.00

e. $103.50

10. State Airlines does a complete maintenance check of its airplanes every 12,000 miles flown. Airplane #200 was flown 96,000 miles last month. How many complete maintenance checks did Airplane #200 have last month?

a. 480 maintenance checks

b. 84,000 miles

c. 108,000 miles

d. 8 maintenance checks

e. 1,152,000,000 miles

11. After getting a tune-up for his SUV, Ernst was able to drive 283.1 miles on 14.9 gallons of gasoline. How many miles did he get per gallon?

a. 19 miles per gallon

b. 42.18 miles per gallon

c. 134 miles per gallon

d. 298 miles per gallon

e. 268.2 miles per gallon

12. The Platte River is normally 7 feet deep. During a recent flood, it crested at 14 feet above normal. What was the depth of the river at the crest of the flood?

a. 7 feet

b. 21 feet

c. 98 feet

d. 2 feet

e. none of the above

13. The supersonic airplane *Concorde* was able to fly 3,855 miles across the Atlantic Ocean in $3\frac{3}{4}$ hours. What was its average speed?

a. $3,858\frac{3}{4}$ miles per hour

b. $3,851\frac{1}{4}$ miles per hour

c. 1,028 miles per hour

d. $14,456\frac{1}{3}$ miles per hour

e. $467\frac{9}{33}$ miles per hour

PERCENT WORD PROBLEMS

Identifying the Parts of a Percent Word Problem

Read the statements below.

The 8-ounce glass is 50% full. It contains 4 ounces.

These statements contain three facts:

the whole: the 8-ounce glass

the part: 4 ounces

the percent: 50%

A one-step percent word problem would be missing one of these facts. When you are solving a percent word problem, first identify what you are looking for. As shown above, you have three possible choices: *the part, the whole, or the percent.*

It is usually easiest to figure out that you are being asked to find the percent. Word problems asking for the percent usually ask for it directly, for example, "What is the percent?" or "Find the percent." or "Three is what percent of . . .?" Occasionally, other percent-type words are used, such as "What is the *interest rate*?"

EXAMPLE 1   **6 is what percent of 30?**

The question asks *is what percent?* Therefore, you are looking for the percent.

Sometimes you are given the percent and one other number. You must decide whether you are looking for the part or the whole.

EXAMPLE 2   **81% of what number is 162?**

The phrase *of what number* means you are looking for the whole.

EXAMPLE 3 Yesterday 114 city employees were absent. This was 4% of the city workforce. How many people work for the city?

STEP 1 *question:* How many people work for the city?

STEP 2 *necessary information:* 114 city employees absent, 4%

STEP 3 You are given the number of city employees who were absent (114) and the percent of the workforce that this represents (4%). You are looking for the total number of people who work for the city, the whole.

EXAMPLE 4 What number is 75% of 40?

You are looking for a number that is a percent of another number. You are looking for the part.

EXAMPLE 5 Operating at full capacity, the automobile plant produced 25 cars an hour. How many cars did the plant produce when operating at 40% capacity?

STEP 1 *question:* How many cars did the plant produce?

STEP 2 *necessary information:* 25 cars an hour, 40% capacity

STEP 3 You are given the production at full capacity (25 cars an hour). To find the production at 40% capacity, you solve for the part.

For each problem, write down whether you are looking for *the part*, *the whole*, or *the percent*. DO NOT SOLVE!

1. The city reported that 14,078 out of 35,817 registered voters voted in the election. What percent of registered voters voted in the election?

2. A total of 14,615 people voted in the election. The election results are shown at the right. How many votes did the winning candidate get?

Vote Percentages	
Candidate A	54%
Candidate B	39%
Candidate C	7%

3. An ad said that 36% of the plumbers polled recommended Drano. If 72 plumbers recommended Drano, how many plumbers were polled?

4. The Phoenix Suns made 85% of 36 free throws in their first game of the season. How many free throws did they make in the game?

5. A seed company guaranteed 87% germination of its spinach seed. If Marco had 450 spinach seeds germinate, how many seeds did he plant?

6. The state had a work force of 1,622,145. If 132,998 of these people were unemployed, what was the unemployment rate for the state?

7. A bedroom set normally priced at $1,400 is on sale. How much would Rochelle save if she bought the set at the advertised sale, shown at the right, instead of at the regular price?

FURNITURE SALE
ALL BEDROOM SETS
40% OFF

8. Last year, 980 people took the high school equivalency exam at the local official test center. If 637 people passed the exam, what percent of the people taking the exam passed?

9. If 8% of the registered voters sign the initiative petition, it will be placed on the November ballot. There are 193,825 registered voters in the county. How many of them must sign the petition for it to go on the ballot?

10. An independent study group estimated that only 35% of all crimes in the city were reported. Last year 2,800 crimes were reported. According to the study, how many crimes were actually committed?

Solving Percent Word Problems

Once you identify what you are looking for in a percent word problem, set up the problem and solve it.

Percent word problems can be solved using proportions. These problems can be set up in the following form:

$$\frac{\text{part}}{\text{whole}} = \frac{n\%}{100}$$

This proportion means that the ratio of the part to the whole is equal to the ratio of the percent to 100.

Using the proportion method, you can solve for the part, the whole, or the percent. The percent is always written over 100 because the percent represents a fraction with 100 in the denominator.

As you saw in your earlier work with proportions, a proportion is the same as two equivalent fractions. For example,

2 is 50% of 4 and can be written as

$$\frac{2}{4} = \frac{50}{100}$$

2 is the *part*, 4 is the *whole*, and 50 is the *percent*.

Remember: There are 100 cents in a dollar. *Percent* means per hundred.

EXAMPLE 1   If 24 out of 96 city playgrounds need major repairs, what percent of the city playgrounds need major repairs?

STEP 1   *question:* What percent of the city playgrounds need major repairs?

You are looking for the percent.

STEP 2   *necessary information:* 24 out of 96

96 is the whole (all the playgrounds).

24 is the part (playgrounds needing repairs).

STEP 3   *numbers*   *percents*

$$\frac{24 \text{ playgrounds}}{96 \text{ playgrounds}} = \frac{n}{100}$$

STEP 4   Cross multiply. Then divide.

$n = \textbf{25\%}$

$$96 \times n = 24 \times 100$$
$$96n = 2{,}400$$
$$n = \frac{2{,}400}{96} = 25\%$$

EXAMPLE 2 The finance company requires that Lynn make a down payment of 15% on a used car. She can afford a down payment of $900. What is the most expensive car that she could buy?

STEP 1 *question:* What is the most expensive car that she could buy?

You are looking for the whole (the price of the car).

STEP 2 *necessary information:* 15%, $900

15% is the percent.

$900 is the down payment, which is a part of the total price of the car.

STEP 3 Set up a proportion.

$$\frac{\$900}{\$n} = \frac{15}{100}$$

STEP 4 Cross multiply. Then divide.

$n =$ **$6,000**

$$15 \times n = 900 \times 100$$
$$15n = 90,000$$
$$n = \frac{90,000}{15} = \$6,000$$

EXAMPLE 3 Jana can spend 35% of her income on rent. She makes $2,740 a month. How much can she spend for rent?

STEP 1 *question:* How much can she spend for rent?

You are looking for the part of her income that she will spend on rent.

STEP 2 *necessary information:* 35%, $2,740

35% is the percent.
$2,740 is her whole income.

STEP 3 Set up a proportion.

$$\frac{\$n}{\$2,740} = \frac{35}{100}$$

STEP 4 Cross multiply. Then divide.

$n =$ **$959**

$$100 \times n = 2,740 \times 35$$
$$100n = 95,900$$
$$n = \frac{95,900}{100} = \$959$$

Use proportions to solve the following problems.

1. A $280 suit was reduced by $84. What was the percent of the reduction?

2. The election results are shown at right. If 28,450 votes were cast in the school board elections, how many votes did Marsha receive?

School Board Elections	
Clayton	19%
Andrea	21%
Marcus	2%
Marsha	58%

3. The state government cut aid for adult education by 25%. Metropolis expects to lose $96,000. How much aid for adult education had Metropolis been receiving?

4. Last year Jeffrey paid 7% of his income in taxes. He paid $2,653. What was his income?

5. Exactly 60% of the residents of the city are African American. The population of the city is 345,780. How many residents are African American?

6. In 1998, Robyn paid $340 interest on $2,000 that she had borrowed. What was the interest rate on the borrowed money?

7. In order to control his cholesterol, Sespend is trying to limit his calories from fat to 20% of his total calories. On an average day, he eats 2,400 calories. What is the most calories from fat he can eat on an average day if he is to meet his goal?

8. According to the nutrition label, a 1-cup serving of Cheerios provides 3 grams of dietary fiber. The label states that this is 11% of the recommended daily amount of fiber. To the nearest gram, how many grams of fiber is the recommended daily amount?

9. Noah and Juno have saved $12,500 to use for a down payment for a house. In order to give them a mortgage, their bank requires the down payment to be at least 5% of the purchase price. What is the highest home purchase price Noah and Juno can afford?

For each word problem, select the correct question that matches the solution.

10. Khan had a special coupon that allowed her to take 20% off the price of a clearance item. She decided to buy a sweater she wanted that had a clearance price of $16.

   a. How much did she pay for the sweater?
   b. How much money did she have left?
   c. How much did she save with her coupon?
   d. What was the original price of the sweater?
   e. How many sweaters could she buy?

   $$\frac{\$n}{\$16} = \frac{20}{100}$$
   $$n = \frac{320}{100} = \$3.20$$

11. A national survey found that 720 out of 900 dentists recommended using Never Break dental floss.

    a. How many dentists recommended a different brand?

    b. How many more dentists recommended Never Break than all other brands?

    c. How many different brands were recommended?

    d. What percent of dentists recommended other brands?

    e. What percent of dentists recommended Never Break?

$$\frac{720 \text{ dentists}}{900 \text{ dentists}} = \frac{n\%}{100}$$

$$900n = 72{,}000$$

$$n = \frac{72{,}000}{900} = 80\%$$

12. The state passed a law requiring 60% of legislators to vote for a tax increase in order for it to pass. The lower house had 180 members.

    a. At least how many legislators in the lower house had to support a tax increase in order for it to pass?

    b. At least how many legislators in the lower house had to oppose a tax increase in order for it not to pass?

    c. What was the largest number of legislators that could oppose a tax increase that passes in the lower house?

    d. How many more legislators are needed to pass a tax increase in the lower house than when only a 50% majority was needed?

    e. What was the smallest possible margin of victory?

$$\frac{n \text{ members}}{180 \text{ members}} = \frac{60}{100}$$

$$100n = 10{,}800$$

$$n = 108 \text{ legislators}$$

13. Floor Mart advertised a total savings of 30% off list price for their store brand microwave oven. When Wilson bought the microwave oven, he saved $75 from the list price.

    a. What percent of the list price did he pay?

    b. What was the list price of the oven?

    c. What was the sale price of the oven?

    d. How much change did Wilson receive?

    e. How much more would he have paid if he had paid full price?

$$\frac{\$75}{\$n} = \frac{30}{100}$$

$$30n = 7{,}500$$

$$n = \frac{7{,}500}{30} = \$250$$

14. The production line for the molded plastic parts needs to be stopped and the machines readjusted if 2% or more of the parts are rejected by the inspectors. In an hour, 600 parts are made on the production line.

    a. What is the largest number of parts that can be defective in an hour?

    b. What is the smallest number of parts that can be defective in an hour?

    c. What is the largest possible number of good parts that can be made in an hour in which the production line needs to be stopped?

    d. What is the smallest possible number of defective parts that need to be found in an hour in order to shut down the production line?

    e. How many more good parts than rejected parts were produced?

$$\frac{n \text{ parts}}{600 \text{ parts}} = \frac{2}{100}$$

$$100n = 1{,}200$$

$$n = 12 \text{ parts}$$

15. In 2006, on a typical day, the United States used approximately 20,590,000 barrels of oil and imported 12,220,000 barrels of oil.

    a. On average, how many barrels of oil were produced in the United States daily?

    b. On average, what percent of daily oil use in the United States is provided by domestic oil production?

    c. On average, what percent of daily oil use in the United States is provided by imported oil?

    d. On average, what is the total of the typical daily use of oil and imported oil for the United States?

    e. On average, what is the difference between the typical daily use of oil and the amount of imported oil in the United States?

$$\frac{12{,}200{,}000 \text{ barrels}}{20{,}590{,}000 \text{ barrels}} = \frac{n\%}{100}$$

$$1{,}220{,}000{,}000 = 20{,}590{,}000n$$

$$\frac{122{,}000}{2{,}059} = n$$

$$n = 59\% \text{ (rounded to the nearest percent)}$$

Percent and Estimation

Have you ever listened to a report of election results? The reporter will often say something like, "The Senator was reelected with 54% of the vote." This percent is an estimate, not an exact amount. The reporter has rounded the result to the nearest percent.

In each situation described, decide whether it is more likely that the percent given is an estimate or an exact amount.

1. The state unemployment rate was 5.3% last month.

   a. exact

   b. estimate

2. During the sale, all clothing prices are reduced by 20%.

   a. exact

   b. estimate

3. Scientists say that there is a 15% chance of a major earthquake in the region in the next ten years.

   a. exact

   b. estimate

4. Even 20 years after the last underground nuclear tests, background radiation levels were 35% above normal.

   a. exact

   b. estimate

5. Angel's son Alberto got a grade of 88% correct on the 50-question multiple-choice test.

   a. exact

   b. estimate

6. In the city, 24% of the people were living in poverty.

   a. exact

   b. estimate

For each word problem, find a solution that could be true. Some questions have a range of possible correct answers. Others have only one possible correct answer.

7. The newscast reported that Elena received 65% of the vote in the local election in which 4,864 votes were cast. How many votes might she have received?

8. The official report said that the city population is expected to increase by 5% in the next decade. The current population is 128,431. What could the population be in a decade and still be within the predicted range?

9. The investment analyst predicted a 12% return on the mutual fund. Sugi invested $3,000 and at the end of the year had earned $352 on that investment. Was the prediction accurate?

10. The state printed 20,000,000 Instant Winner Lottery Tickets and claimed that 3% of the tickets were winners, with the prizes ranging from a free ticket to $100,000. How many winning tickets were printed?

11. A microwave oven manufacturer predicted that its number of microwave oven sales would increase 23% over sales from the previous year. The previous year's total sales were 568,200 microwave ovens. The manufacturer just announced that this year it sold 699,650 microwave ovens. Was the manufacturer's prediction accurate?

The Percent Circle

If you find working with proportions to solve percent problems too abstract, you could use a memory aid called the **percent circle.** The percent circle is equivalent to a proportion, but it creates a picture to help you decide whether to multiply or divide.

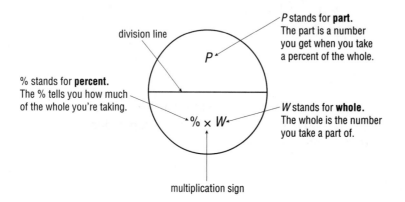

division line

P stands for **part.**
The part is a number you get when you take a percent of the whole.

% stands for **percent.**
The % tells you how much of the whole you're taking.

W stands for **whole.**
The whole is the number you take a part of.

multiplication sign

Using the Percent Circle

EXAMPLE 1 Finding a part of the whole.

According to Francisco's union contract, he is due to get a 4% raise. He currently earns $850 a week. What will be his raise?

STEP 1 The raise is calculated as a part of his salary. Therefore cover *P* (the part) since that is the number you're trying to find.

STEP 2 Read the uncovered symbols: % × *W*.

To find the part, multiply the percent by the whole.

STEP 3 Write the percent as either a fraction or a decimal.

$4\% = \frac{4}{100}$ or 0.04

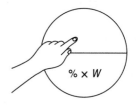

STEP 4 Do the calculation.

$\frac{4}{100} \times 850 = \frac{3,400}{100} = \$34.00$ or $0.04 \times 850 = \$34.00$

Francisco will get a **$34 raise.**

Note: When you use the percent circle, you must remember that *percent* means "compared to 100." When you know the percent, remember to divide it by 100. When you are looking for the percent, remember to multiply by 100.

EXAMPLE 2 **Finding a whole when a part and a percent are given.**

The Cheap Cars used car lot requires buyers to pay 20% of the total price of a car as a down payment. The buyer can finance the rest of the purchase price. Celso can spend up to $800 on a down payment for a used car. What is the highest price used car he could buy at Cheap Cars?

STEP 1 The down payment is a part of the price of a car, so you are trying to find the whole. Therefore, cover W (the whole) since that is the number you're trying to find.

STEP 2 Read the uncovered symbols. $\frac{P}{\%}$ means P divided by %.

To find the whole, divide the part by the percent.

STEP 3 Write the percent as either a fraction or a decimal.

$20\% = \frac{20}{100} = \frac{1}{5}$ or 0.20

STEP 4 Do the calculation.

$800 \div \frac{1}{5} = 800 \times \frac{5}{1} = \$4,000$ or $.20\overline{)800.00}$ = $\$4,000$

The highest price used car Celso could buy at Cheap Cars would cost **$4,000.**

EXAMPLE 3 **Finding what percent a part is of a whole.**

A clinic tested 360 drug abusers for HIV and found that 234 tested positive. What percent of the drug abusers tested HIV positive?

STEP 1 You need to find what percent 234 is of 360.

You are looking for the percent. Therefore, cover %.

STEP 2 Read the uncovered symbols. $\frac{P}{W}$ means P divided by W.

To find the percent, divide the part by the whole.

STEP 3 Do the calculation.

$\frac{234}{360}$ or $360\overline{)234.00}$ = 0.65

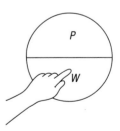

STEP 4 Convert the decimal to a percent.

$0.65 \times 100 = 65\%$

65% of the drug abusers tested HIV positive.

Note: It is possible for the part to be larger than the whole. If the part is larger than the whole, the percent will be larger than 100%.

Use the percent circle to solve the following word problems.

1. Dinner cost Luis $40. He wants to leave the waitress a 15% tip. How much should he leave for a tip?

2. Armand bought a chain saw for $80. How much did he pay for sales tax if the sales tax was 5%?

3. Webmasters Computer Company has 540 employees. Of those employees, 27 are over age 60. What percent of the employees are over age 60?

4. Super Foods' market research showed that 2% of all households that received a weekly circular actually came to shop at the store. The store manager of a new Super Foods planned for 1,500 different customers to visit the Super Foods store in its first week. How many circulars should she plan to send out?

5. Ninety days after the end of training, 24 out of 30 graduates of the computer repair training program found jobs. What was the job placement rate of the computer repair training program?

6. The Serious Disease Charity spent $28,000 to raise $112,000 in contributions. What percent of the money raised was spent on fund-raising?

Solving Percent Word Problems with Decimals and Fractions

Many percent word problems also contain decimals or fractions. These problems are also solved using the proportion method.

EXAMPLE 1 For the big sale, every item in the store was $33\frac{1}{3}\%$ off the marked price. How much would Luc have saved on pants marked $54?

STEP 1 *question:* How much would Luc have saved on pants marked $54?

You are looking for the part.

STEP 2 *necessary information:* $33\frac{1}{3}\%$, $54

$33\frac{1}{3}$ is the percent.

$54 is the whole.

STEP 3 *proportion:*

$$\frac{n}{54} = \frac{33\frac{1}{3}}{100}$$

STEP 4 Cross multiply. Then divide.

Luc would have saved **$18.**

$$100 \times n = 33\frac{1}{3} \times 54$$
$$100n = 1{,}800$$
$$n = \frac{1{,}800}{100} = \$18$$

EXAMPLE 2 Bob takes home $156.40 out of his weekly pay of $184. What percent of his pay does he take home?

STEP 1 *question:* What percent of his pay does he take home?

You are looking for the percent.

STEP 2 *necessary information:* $156.40, $184

$156.40 is the part.

$184.00 is the whole.

STEP 3 *proportion:*

$$\frac{\$156.40}{\$184.00} = \frac{n}{100}$$

STEP 4 Cross multiply. Then divide.

Bob takes home **85%** of his pay.

$$184 \times n = 156.40 \times 100$$
$$184n = 15{,}640$$
$$n = \frac{15{,}640}{184} = 85\%$$

You can also use the percent circle to solve percent word problems with fractions or decimals.

EXAMPLE 3   Specialty Hardware needs to collect 6% sales tax on all items sold. In one day the store collected $424.80 in sales tax. What was the store's total sales for the day?

STEP 1   You are looking for the total sales for the day. You know the taxes collected and the tax rate, so you are looking for the whole. Therefore, cover W (the whole) since that is the number you're trying to find.

STEP 2   Use the percent circle. $\frac{P}{\%}$ means P divided by %.

To find the whole, divide the part by the percent.

STEP 3   Convert the percent to a fraction or a decimal.

$6\% = \frac{6}{100}$ or 0.06

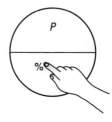

STEP 4   Do the calculation.

$$424.80 \div \frac{6}{100} = 424.80 \times \frac{100}{6} = \frac{42,480}{6} = \$7,080 \quad \text{or} \quad .06\overline{)424.80} = \$7,080$$

The total sales for the day was **$7,080.**

Solve the following percent problems using proportions or the percent circle.

1. The local newspaper reported that 90 homes were damaged in the flood. The report also stated that this was 4.5% of all the homes in the town. Approximately how many homes were in the town?

2. Tanya trimmed $\frac{1}{10}$ pound of fat off $\frac{3}{4}$ pound of boneless chicken breast. What percent of the total weight of the chicken did she trim?

3. According to the Web site's statistical tracker, 6.4% of 800 visitors to the Web site yesterday were visiting for the first time. How many visitors to the Web site were visiting for the first time yesterday?

4. Russo's Restaurant collected $49.76 in taxes Friday night. The food tax is 8%. How much money did the restaurant receive for meals on Friday night?

For problems 5–7, use the tax guidelines chart shown at the right.

5. Jed bought a sandwich for $8.60. Find the amount of tax he paid.

State Tax Guidelines
Clothing—6%
Food—5%
Alcohol—7%
Cigarettes—8%

6. Sylvia bought a skirt for $42.50. Find the amount of tax she paid.

7. Flavia bought a bottle of wine that cost $15.95 for a dinner party. Find the amount of tax she paid.

8. Glenda bought maple syrup for $7.92 a pint and sold the syrup for $3.96 a pint more. By what percent did she mark up the price of the maple syrup?

9. Barry's Bookstore is having a storewide book sale in which all prices are cut by $12\frac{1}{2}$%. How much did Juan save on a book that normally costs $12.80?

10. A telemarketing company found that 3.5% of its calls resulted in new subscribers. If its target is 12,000 new subscribers, how many calls should the company plan to make?

Solving More Percent Word Problems

The following problems give you a chance to review percent word problems containing whole numbers, decimals, and fractions.

Solve the problems by using proportions or the percent circle.

1. Of the dentists surveyed, 3 out of 4 recommend a fluoride toothpaste. What percent of the dentists surveyed recommend a fluoride toothpaste?

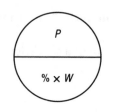

2. Of the registered voters, 112,492 people voted for mayor in the city. This was 40% of the registered voters. How many registered voters are there in the city?

3. In a normal season, the Seaside Resort has 34,500 visitors. This year, due to bad weather, 11,500 fewer visitors came to the resort. What was the percent drop in business for the resort?

4. The High Tech Electronics Company announced an 8.6% profit on sales of $49,600,000. How much profit did the company make?

5. The Quality Chocolate Company decided to increase the size of its chocolate bar by 0.4 ounce. This was an increase in size of $16\frac{2}{3}$%. What was the weight of its chocolate bar before the change?

6. In a recent flu epidemic, 0.8% of people over age 65 who caught the disease died. The death toll in this group was 40. How many people over age 65 caught the flu?

7. Last year, Dennis paid 13% of his income in taxes. He earned $41,694. How much did he pay in income taxes?

8. The Housekeeper's Union has just won a 7% raise for its members. Estella is a union member currently making $17,548. How much of a raise will she get?

9. Nayana received a 9% raise worth $18 a week. What had her week's salary been?

10. Basketball champ Kareem scored on 506 out of 1,012 attempts. What was his scoring percentage?

11. A $\frac{3}{4}$-cup serving of Honey Nut Cheerios served with skim milk provides 30% of the U.S. recommended daily allowance of vitamin A. How many cups of Honey Nut Cheerios should you eat in order to receive the full U.S. recommended daily allowance of vitamin A when you have skim milk with each serving?

12. A $3\frac{3}{4}$-ounce serving of Norway Sardines in chili sauce provides 100% of the U.S. recommended daily allowance for vitamin D. What percent of the U.S. daily allowance is provided per ounce of the Norway Sardines in chili sauce?

Solving Combination Word Problems

Until now, this book has shown you word problems that can be solved in one step or using a proportion. However, many situations require you to use a combination of math operations to solve word problems.

Generally, you can solve these combination problems by breaking them into two or more one-step problems. As you read a word problem, you may see that it will take more than one math operation to solve. The difficulty lies in deciding how many steps to take and in what order to work them out.

The key to solving combination problems is—start with the question and work backward.

This shouldn't be difficult. Throughout this book, you have started your work by finding the question.

Follow these steps in solving combination word problems.

STEP 1   Find the question.

STEP 2   Select the necessary information.

STEP 3   Write a solution sentence for the problem. Fill in only the necessary information that belongs in the solution sentence.

Write another sentence, this time to find the information that is missing in the solution sentence. Solve the sentence that gives you the missing information.

STEP 4   Fill in the missing information (the answer from Step 3) in the solution sentence and solve.

STEP 5   Make sure that the answer is sensible.

No matter how many short problems a combination problem consists of, you can always work backward from the solution sentence. Examples 1 and 2 illustrate this.

EXAMPLE 1 Sengchen had $38 in her checking account. She wrote checks for $14 and $9. How much money was left in her account?

STEP 1 *question:* How much money was left in her account?

STEP 2 *necessary information:* $38, $14, $9

STEP 3 Write a solution sentence.

money − checks = money left

Fill in the sentence with information that can be used to solve the problem.

$$\$38 - \text{checks} = \text{money left}$$

Decide what *missing information* is needed to solve the problem. Write a number sentence and solve.

check + check = checks

$$\$14 + \$9 = \boxed{\$23}$$

$$\$38 - \boxed{\$23} = \text{money left}$$

Now you have the complete information needed to solve the problem.

STEP 4 Solve.

$$\$38 - \$23 = \textbf{\$15 left}$$

$15 was left in her account.

EXAMPLE 2 Lillie is a travel agent. She arranges a trip for 56 people at a cost of $165 for airfare plus $230 for hotel per person. How much money does she collect from the group?

STEP 1 *question:* How much money does she collect?

STEP 2 *necessary information:* 56 people, $165 airfare, $230 hotel accommodations

STEP 3 Write a solution sentence.

cost × number of people = total

$$\text{cost} \times 56 = \text{total}$$

Solve for *missing information*.

airfare + hotel = cost

$$\$165 + \$230 = \boxed{\$395}$$

$$\boxed{\$395} \times 56 = \text{total cost}$$

STEP 4 Solve.

$$\$395 \times 56 = \textbf{\$22,120 total}$$

Lillie collects **$22,120** from the group.

The words that you use in the solution and missing information sentences may differ from what is presented here. What is important is that you break down the problem into smaller steps.

A combination word problem that needs both multiplication and division can often be written as one proportion instead of two separate word sentences.

EXAMPLE 3 The posted price for apples was $1.58 for 2 pounds. The Manager's Special offered an additional 35 cents off if you buy 3 pounds of apples. How much did Michelle pay for 3 pounds of apples?

STEP 1 *question:* How much did Michelle pay for 3 pounds of apples?

STEP 2 *necessary information:* 2 pounds, $1.58, 3 pounds, additional 35 cents off

$$\frac{pounds}{cents} = \frac{pounds}{cents}$$

STEP 3 Write a solution sentence:

Regular price of 3 pounds − discount = sale price

To find the regular price, write a proportion to show the relationship between weight and cost.

$$\frac{2\ pounds}{\$1.58} = \frac{3\ pounds}{n}$$

$$2 \times n = 3 \times 1.58$$

$$n = \frac{\$4.74}{2} = \$2.37$$

STEP 4 Fill in the appropriate numbers and solve.

Michelle paid **$2.02** for 3 pounds of apples.

$$\$2.37 - \$0.35 = \$2.02$$

For each word problem, write two word sentences (a solution sentence and a missing information sentence) or a proportion. DO NOT SOLVE!

1. Tim earns $880 a week. Every week $249 in taxes and $26 in union dues are taken out of his paycheck. What is his take-home pay?

2. After starting the day with $71, Miguel spent $5 for lunch and $32 for gasoline. How much money did he have left by the end of the day?

3. A store bought 30 boxes of dolls for $7,200. If there were 8 dolls in a box, how much did each doll cost?

4. Samuel had $394 in his checking account. After he wrote a check for $187 and deposited $201, how much money was in his checking account?

5. Kelly bought five of the shirts shown at the right and one skirt. How much money did she spend on these items?

$19

6. Martha borrows $4,600 to buy a new car. Over the life of the loan, she will have to pay $728 interest. She plans to pay back the loan plus the interest in 24 equal monthly payments. How much will her monthly payments be?

$24

7. To be hired, a data entry operator must be able to enter numbers into a computer at the rate of 10,000 numbers every 60 minutes. Yvana took a 15-minute data entry test. How many numbers did she have to enter to be hired?

8. A 14-gram serving of Cains Mayonnaise contains 5 grams of polyunsaturated fat and 2 grams of saturated fat. How many grams of saturated fat are in a 224-gram jar of Cain's Mayonnaise?

9. At Burger Queen, a cheeseburger costs $2.89, medium french fries cost $1.79, and a medium cola costs $1.49. The cheeseburger combination meal of a cheeseburger, medium fries, and medium cola costs $5.29. How much does Olga save by buying the cheeseburger combination meal instead of buying the three items separately?

Write two one-step word sentences or a proportion needed to solve the following combination word problems. Then solve the problems.

10. For the convention, each of the 8 wards of the city elected 4 delegates, while 5 delegates were elected at-large. How many delegates did the city send to the convention?

11. It cost the gas station owner $181 in parts and $245 in labor to fix his customer's car. He charged his customer $563. How much profit did the owner make on the job?

12. Mark was offered a job downtown that would give him a raise of $78 a month over his current salary, but his commuting costs would be $2-a-day higher. If he works 22 days a month, what would be his net monthly increase in pay?

13. The recipe at the right serves 6 people. Ginny is planning to make the recipe for 8 people. How much stew beef does she need?

Hearty Stew
3 cups beef stock
1 cup red wine
$\frac{1}{4}$ cup oil
2 pounds potatoes
3 pounds stew beef
1 pound carrots

14. Last month Elvira was billed $90 for using 720 kilowatts of electricity. This month she checked her meter and found that she had used 648 kilowatts of electricity. Assuming the cost of electricity has not changed, how much less will her electric bill be for this month?

15. Carlos has a large landscaping job that requires 1,728 cubic feet of loam. His pickup truck can carry 48 cubic feet of loam. He can deliver 6 loads of loam each workday. How many workdays will it take Carlos to deliver all the loam he needs for his landscaping job?

Solving Problems with Decimals, Fractions, and Percents

Decimal, fraction, and percent combination word problems are set up and solved in the same way as whole-number combination word problems.

EXAMPLE 1   For each child at her daughter's birthday party, Shelly spent $2.69 for a party favor and $1.49 for a balloon. There were 13 children in all. How much did Shelly spend on gifts for the children?

STEP 1   *question:* How much did Shelly spend on gifts for the children?

STEP 2   *necessary information:* $2.69, $1.49, 13 children

STEP 3   *solution sentences:*

gifts × children = total cost $gifts \times 13 = total\ cost$

missing information sentence:

favor + balloon = gifts

$2.69 + $1.49 = $4.18

STEP 4   Solve. $\$4.18 \times 13 = total\ cost$

Shelley spent **$54.34** on gifts for the children. $\$4.18 \times 13 = \$54.34$

EXAMPLE 2   Bright's department store advertised that everything in the store was $\frac{1}{5}$ off. Debra bought a pair of pants priced at $40. How much did the pants cost after the discount?

STEP 1   *question:* How much did the pants cost after the discount?

STEP 2   *necessary information:* $\frac{1}{5}$, $40

STEP 3   *solution sentence:*

original price − discount = sale price $\$40 - discount = sale\ price$

missing information sentence:

price × fraction = discount

$\$40 \times \frac{1}{5} = \$8$

STEP 4   Solve. $\$40 - \$8 = sale\ price$

The pants cost **$32** after the discount. $\$40 - \$8 = \$32$

Both of these examples illustrate two-step word problems. Later in this chapter you will work with problems that need more than two steps for solution.

Real Value Hardware advertised that all prices have been reduced 15%. A seven-piece screwdriver set is on sale for $13.60. What was its original price?

STEP 1 *question:* What was its original price?

STEP 2 *necessary information:* 15%, $13.60

STEP 3 *solution statement:* Since this is a percent problem, you can write a proportion.

$$\frac{\text{sale price}}{\text{original price}} = \frac{\text{percent sale}}{100}$$

$$\frac{13.60}{n} = \frac{\text{percent sale}}{100}$$

missing information:

$$100\% - \text{percent reduced} = \text{percent sale}$$

$$100\% - 15\% = 85\%$$

$$\frac{13.60}{n} = \frac{85}{100}$$

$$85 \times n = 13.60 \times 100$$

STEP 4 Solve.

The original price was **$16.**

$$85n = 1{,}360$$

$$n = \$16$$

Write the word sentences and/or proportion needed to solve the following combination word problems. Then solve the problems.

1. Chris bought 6 boxes of cookies for $24.40. If there were 20 cookies in a box, how much did each cookie cost? (Round to the nearest penny.)

2. The $400 washing machine at the right was reduced for a clearance sale. What was its clearance price?

3. Susan bought five cans of pears, each containing $9\frac{3}{4}$ ounces of pears, and one can of fruit cocktail containing $17\frac{1}{2}$ ounces of fruit. What was the total weight of the fruit she bought?

4. Phil bought a water widget that cost $2.49. It reduced his hot water use and saved him $3.40 a month in costs. What were his net savings for 12 months?

5. When cooked, a hamburger loses $\frac{1}{3}$ of its original weight. How much does a $\frac{1}{4}$-pound hamburger weigh after it is cooked?

6. The tax on a meal is 6%. How much is Milton's total bill on a $24 dinner?

7. During the summer clearance sale, everything in the store was 30% off. Solaire bought the bathing suit advertised at the right. How much did she pay for the suit?

30% Off the Prices Below!
Shorts—regularly $20.50
Shirts—regularly $19.00
Socks—regularly $3.99
Bathing Suits—regularly $39.50

8. Marlene bought a new couch for $610.60. She paid $130 down and planned to pay the rest in 12 equal monthly payments. How much will she pay each month?

9. Walter bought a case of 30 bottles of cooking oil for $57.00. He then sold the oil for $2.10 per bottle. How much money did he make on each bottle?

10. Harry bought a case of 30 bottles of cooking oil for $57.00. He then sold the oil for a profit of $0.20 per bottle. What was the percent of profit? (Round to the nearest tenth of a percent.)

11. At her diner, Mireya added 0.2 ounce of salt to a 4-gallon (256 ounce) pot of hearty beef stew. How much salt was in each 12-ounce portion? (Round to the nearest hundredth.)

12. A clothing manufacturer makes a top and matching skirt. The manufacturer buys material for the clothes in 60-inch-wide rolls. Each skirt pattern requires $\frac{2}{3}$ yard of material, while each top requires $1\frac{1}{4}$ yards of material. How many complete outfits can be made from a 70-yard long roll of material?

Order of Operations

You've seen how to use solution sentences to solve combination word problems. If you know and use the rules for order of operations, you can write out the steps of a multistep word problem in one line.

Following are the rules for order of operations in arithmetic expressions.

Rule 1: Do multiplication and division before addition and subtraction.

EXAMPLE 1 $9 - 2 \times 4$

SOLUTION: Multiply. $2 \times 4 = 8$

Subtract. $9 - 8 = $ **1**

EXAMPLE 2 $24 \div 4 + 2$

SOLUTION: Divide. $24 \div 4 = 6$

Add. $6 + 2 = $ **8**

Rule 2: Do the arithmetic inside parentheses first.

EXAMPLE 1 $(21 - 6) \div 3$

SOLUTION: Subtract. $21 - 6 = 15$

Divide. $15 \div 3 = $ **5**

EXAMPLE 2 $5 \times (4 + 8)$

SOLUTION: Add. $4 + 8 = 12$

Multiply. $5 \times 12 = $ **60**

Rule 3: Using Rules 1 and 2, evaluate arithmetic expressions starting from the left.

EXAMPLE 1 $20 - 6 + 4$

SOLUTION: Subtract. $20 - 6 = 14$

Add. $14 + 4 = \mathbf{18}$

EXAMPLE 2 $1 + 2 \times 9 \div 6$

SOLUTION: Multiply. $2 \times 9 = 18$

Divide. $18 \div 6 = 3$

Add. $1 + 3 = \mathbf{4}$

Using the rules for order of operations, evaluate the following arithmetic expressions.

1. $(5 \times 4) - (6 + 3)$

2. $5 \times 4 - 6 + 3$

3. $145.6 + 12.2 - 5.7 - 1.1$

4. $3.2 \times 6 + 7.8$

5. $(7.8 + 2.2) \div 5$

6. $(4 \times 9) \div 3 + 6$

7. $4 \times 9 \div 3 + 6$

8. $1.8 \div 0.02 - 10 - 8$

9. $1.8 \div 0.02 - (10 - 8)$

10. $56 - 7 \times 3.5$

Using Order of Operations in Combination Word Problems

Look at the following examples to see how to use your knowledge of order of operations to set up a multistep word problem.

EXAMPLE 1   Frank bought dinner for two for $22.50. His bill included a 5% meal tax. What was his total bill?

STEP 1   *question:* What was his total bill?

STEP 2   *solution sentence:*

price of dinner + (5% × price of dinner) = total bill

STEP 3   *set up:*

$22.50 + (0.05 × 22.50) = total bill

STEP 4   *solution:*

Multiply. 0.05 × $22.50 = $1.125

$\qquad\qquad\qquad$ = $1.13 (nearest cent)

Add. $22.50 + $1.13 = **$23.63**

Frank's total bill was **$23.63.**

EXAMPLE 2   Edna lives 19 miles from work. Last week she drove to work and back 5 days and on the weekend drove 170 miles to visit relatives. How many miles did she drive last week commuting and visiting relatives?

STEP 1   *question:* How many miles did she drive?

STEP 2   *solution sentence:*

commuting + visiting relatives = total miles

STEP 3   *set up:*

(19 miles × 5 days × 2 times a day) + 170 miles = total miles

STEP 4   *solution:*

Multiply. 19 × 5 × 2 = 190 miles

Add. 190 miles + 170 miles = **360 miles**

Edna drove a total of **360 miles.**

For each word problem, circle the letter of the correct setup.

1. Three friends went out to dinner. Their meal cost $56.88. They left a tip of $9.00. How much was each person's share if they divided the total amount equally?

   a. ($56.88 ÷ 3) + $9.00

   b. $56.88 + $9.00 ÷ 3

   c. $56.88 + ($9.00 × 3)

   d. ($56.88 + $9.00) ÷ 3

   e. ($56.88 + $9.00) × 3

2. Kathy and Peter went clothes shopping for their baby daughter, Gina. They bought five sleepers for $9.95 each and six T-shirts for $3.40 each. How much money did they spend?

   a. 5 × $9.95 + 6 × $3.40

   b. (5 + $9.95) × (6 + $3.40)

   c. (5 + 6) × ($9.95 + $3.40)

   d. ($9.95 ÷ 5) + ($3.40 × 6)

   e. ($9.95 + $3.40) ÷ (5 + 6)

3. Katrina paid $99.90 for her monthly train pass. If she used her pass twice a day for 23 days last month, what was the average cost of each ride?

   a. ($99.90 ÷ 23) × 2

   b. $99.90 ÷ (23 × 2)

   c. $99.90 − (23 × 2)

   d. $99.90 ÷ 23

   e. $99.90 ÷ 23 × 2

4. A coat normally selling for $120 was marked down 40%. What was the sale price?

   a. $120 − 40

   b. $120 ÷ 40

   c. $120 − 0.40 × $120

   d. $120 + 0.40 × $120

   e. $120 − $120 ÷ 40

5. Angela needs to buy eggs for the week. She will make six 3-egg omelets for her family. She will also need 4 eggs for a cake and 2 eggs for a casserole. How many dozen eggs must she buy?

   a. (6 × 3) + 4 + 2

   b. (6 + 3 + 4 + 2) ÷ 12

   c. 12 × (6 ÷ 3) + (4 ÷ 2)

   d. 6 × 3 ÷ 12 + 4 + 2 ÷ 12

   e. (6 × 3 + 4 + 2) ÷ 12

Using Memory Keys for Multistep Word Problems

Now that you know the order of operations, you can use the memory keys on your calculator for multistep word problems.

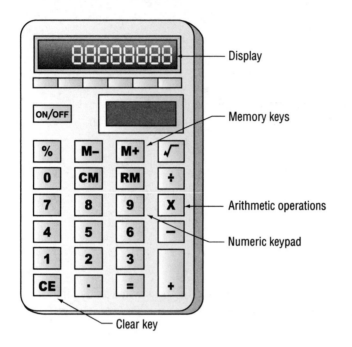

First you need to write out your setup of the solution of a word problem. Then following the order of operations, do each calculation. You can either add or subtract the result to your calculator memory when you need to. The memory stores these results until it is cleared with either a Clear All or a Clear Memory.

There are some multistep calculations that do not need the memory keys at all, such as a series of multiplications. Practice with your calculator to see if you can find more than one correct procedure for solving a multistep word problem.

The following is an example of how you could use the memory keys to solve a multistep word problem. Since not all calculators handle memory the same way, work through this example on your calculator to make sure these directions work on it.

EXAMPLE   For a party, Hanan bought six bottles of soda for $1.49 each and three bags of chips for $2.79 each. How much did she spend?

STEP 1   Set up the solution.

cost of soda + cost of chips = total cost

$(6 \times 1.49) + (3 \times 2.79)$ = total cost

STEP 2 Key in the first operation in parentheses and add it to memory.

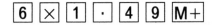

(You may need to use the = key before the memory key.)

You should see 8.94 in your display.

STEP 3 Key in the second operation in parentheses and add it to the amount already in memory.

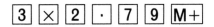

(You may need to use the = key before the memory key.)

You should see 8.37 in your display.

STEP 4 Press RM (sometimes labeled MRC or RCM) to read the result. You should see 17.31 on the display.

STEP 5 Press CM (Clear Memory) or CA (Clear All) to clear the memory before you start calculating another problem.

For each word problem, look at the solution setup and select the correct question. Then use the setup and your calculator to solve the problem.

1. Suki has $46 left on her debit card and $48 in cash. She bought $81 in groceries.

   ($46 + $48) − $81

   a. How much more did Suki have in cash?

   b. How much money did Suki have left?

   c. How much money did she have before she paid for her groceries?

   d. If she had forgotten her debit card, how many additional dollars would she have needed to pay her bill?

   e. How much money does Suki need to borrow to pay her bill?

2. Emily mixes a fruit cooler with 5 fluid ounces of fruit juice to 3 fluid ounces of seltzer. For a party she wants to make 2 gallons (256 fluid ounces) of fruit cooler.

$$\frac{5}{5+3} = \frac{n}{256}$$

   a. How much seltzer will she need?

   b. How much more fruit juice than seltzer will she use?

   c. What is the size of one serving?

   d. How many servings is she making?

   e. How much fruit juice will she need?

3. The clearance advertisement said that an additional 20% would be taken off the labeled price at the register. Nerys bought a pair of boots marked $85 and a belt marked $25.

$$\frac{n}{\$85} = \frac{100 - 20}{100}$$

   a. How much did Nerys save on the boots?

   b. How much did Nerys pay for the boots?

   c. What percent of the original price did Nerys have to pay?

   d. How much did Nerys save on the two items?

   e. How much did Nerys save on the belt?

4. Ultra Clean detergent comes in two sizes. The 64-fluid ounce size costs $11.98. The 32-fluid ounce size costs $6.79.

($6.79 ÷ 32) − ($11.98 ÷ 64)

   a. How much more does the larger size cost?

   b. How much less does the larger size cost per fluid ounce?

   c. What is the per fluid ounce cost of the larger size?

   d. What is the per fluid ounce cost of the smaller size?

   e. How much more detergent is in the larger size?

Using Pictures or Diagrams to Set Up Multistep Word Problems

You can use pictures or diagrams to help you set up and solve multistep word problems.

EXAMPLE 1 Holly just moved into her new studio apartment. The main room is 30 feet long by 15 feet wide, and the bathroom is 9 feet long by 8 feet wide. What is the total square footage of her apartment?

STEP 1 *question:* What is the total square footage of her apartment?

STEP 2 *picture:*

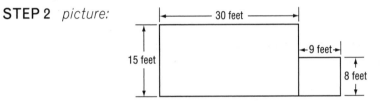

STEP 3 *set up:*

The square footage of the apartment is the sum of the area of each room. Each room is the shape of a rectangle. The area of a rectangular room is found by multiplying its length by its width.

area of main room + area of bathroom = total square footage

(30 feet × 15 feet) + (9 feet × 8 feet) = total square footage

STEP 4 *solution:*

Multiply. 30 feet × 15 feet = 450 square feet

Multiply. 9 feet × 8 feet = 72 square feet

Add. 450 square feet + 72 square feet = **522 square feet**

EXAMPLE 2 Donna's car gets 27 miles per gallon. After filling her 18-gallon gasoline tank, she drove 135 miles. How many more miles can she drive before her car runs out of gasoline?

STEP 1 *question:* How many more miles?

STEP 2 *diagram:*

```
|←————————— 27 × 18 miles —————————→|
|————————————|——————————————————————|
   135 miles        ? miles left
```

STEP 3 *set up:*

(27 mpg × 18 gallons) − 135 miles = miles left

STEP 4 *solution:*

Multiply. 27 mpg × 18 gallons = 486 miles

Subtract. 486 miles − 135 miles = **351 miles**

For each word problem, draw a picture or diagram and circle the letter of the correct setup.

1. A set of silverware has 8 place settings. It includes a knife, dinner fork, salad fork, teaspoon, and soup spoon, as well as 6 serving pieces. How many pieces of silverware are in the set:

   a. 8×6

   b. $8 \times 6 - 5$

   c. $8 \times 5 + 6$

   d. $5 + 8 \times 6$

   e. $8 \times 6 \div 5$

   diagram:

2. A florist has 420 roses and 380 carnations. How many bouquets, each containing 5 flowers, can be made using these flowers?

   a. $(420 + 380) \times 5$

   b. $5 \times 420 - 380$

   c. $5 \div (420 + 380)$

   d. $420 \times 5 - 380 \times 5$

   e. $420 \div 5 + 380 \div 5$

   diagram:

3. Jan turns a 9-foot by 12-foot room in her home into an office. If she wants three equally sized bookcases along one of the 9-foot walls, how wide can each bookcase be?

   a. $(9 \div 3) \times 2$

   b. $(9 \times 12) \div 3$

   c. $(9 \div 3) \times 12$

   d. $9 \div 3$

   e. $(12 \div 3) \times 9$

   diagram:

4. The theater has 30 rows of 25 seats each. If 150 of the seats are taken, how many are empty?

   a. $(30 \times 25) - 150$

   b. $(30 \times 25) \div 150$

   c. $150 - 30 \times 25$

   d. $(30 + 25) \times 150$

   e. $(150 \div 30) + 25$

   diagram:

For each word problem, draw a picture or diagram and circle the letter of the correct setup.

5. Tanya and Scott had lunch at the local fast-food restaurant: Tanya had a 680-calorie hamburger, a 90-calorie salad, and 0-calorie bottled water. Scott had a 560-calorie chicken sandwich, a 600-calorie order of french fries, and a 660-calorie strawberry milk shake. How many more calories was Scott's meal?

   a. $560 + 600 + 660 + 680 + 90$ diagram:

   b. $560 + 600 + 660 - 680 - 90$

   c. $560 + 600 - 660 - 680 - 90$

   d. $(560 + 600 + 660) \times (680 + 90)$

   e. $(560 + 600 + 660) - (680 - 90)$

6. On average, the Second Home Kennel feeds each of the dogs it boards 0.7 pound of dog food a day. The manager buys the dog food in 17.5-pound bags. If Second Home has 36 dogs in the kennel, how many bags of dog food are needed for one day?

   a. $36 - (17.5 \times 0.7)$ diagram:

   b. 0.7×36

   c. $36 - 17.5$

   d. $\dfrac{0.7 \times 36}{17.5}$

   e. $0.7 \times 36 \times 17.5$

7. Magda is a housekeeper at the Deluxe Hotel. She puts fresh towels in each room every day. A regular room needs 2 large towels, 3 medium towels, 2 small towels, and 2 washcloths. The suites get 4 large towels, 4 medium towels, 4 small towels, and 4 washcloths. If her floor has 8 regular rooms and 1 suite, how many towels should Magda load in her cart to put in the rooms?

   a. $2 + 3 + 2 + 2 + 4 \times 4$ diagram:

   b. $(2 + 3 + 2 + 2) \times 8 + 4 \times 4$

   c. $2 + 3 + 2 + 4 \times 3$

   d. $(2 + 3 + 2 + 2) \times 8 - 4 \times 3$

   e. $(2 + 3 + 2 + 2) \times 8 + 4 \times 3$

8. The Flower Greenhouse sells flowers in $\frac{1}{2}$-ft by $\frac{2}{3}$-ft flats. The flats are placed on 12-ft by 4-ft display tables. There are 24 display tables in all. What is the greatest number of flats that can be displayed at one time?

   a. $(12 \times 4) \div (\frac{1}{2} \times \frac{2}{3}) \times 24$ diagram:

   b. $(12 \times 4) \times (\frac{1}{2} \times \frac{2}{3}) \times 24$

   c. $(12 \times 4) \div (\frac{1}{2} \times \frac{2}{3}) \div 24$

   d. $(12 \times 4) \times (\frac{1}{2} \times \frac{2}{3}) \div 24$

   e. $(12 \div 4) \times (\frac{1}{2} \times \frac{2}{3}) \times 24$

Solving Combination Word Problems Involving Conversions

Many combination word problems involve conversions. To solve the following examples and problems, you should refer to the conversion chart on page 221. You should notice that a conversion is needed when one unit of measurement appears in the necessary information and a different unit of measurement is called for in the question.

EXAMPLE 1   A dairy farm sold 156 *quarts* of milk at its own store and shipped out an additional 868 *quarts* to nearby supermarkets. How many *gallons* of milk were marketed?

STEP 1   *question:* How many gallons of milk were marketed?

STEP 2   *necessary information:* 156 quarts, 868 quarts

STEP 3   *solution statement:*

$$\frac{\text{quarts}}{1\ \text{gallon}} = \frac{\text{total quarts}}{\text{total gallons}}$$

$$\frac{4}{1} = \frac{quarts}{gallons}$$

There are 4 quarts in a gallon. This is written on the left side of the proportion as $\frac{4}{1}$.

$$\frac{4}{1} = \frac{1,024}{n}$$

missing information sentence:

quarts + quarts = total quarts

156 + 868 = 1,024 quarts

$$4 \times n = 1 \times 1,024$$

$$4n = 1,024$$

STEP 4   Solve.

$$n = \frac{1,024}{4} = 256\ gallons$$

EXAMPLE 2   A mill is cutting 8-*foot* lengths of lumber into chair legs. There are 6 *inches* of scrap for each length. What percent of the wood is scrap?

STEP 1   *question:* What percent of the wood is scrap?

STEP 2   *necessary information:* 8-foot lengths, 6 inches

STEP 3   *solution statement:*

$$\frac{\text{scrap}}{\text{length of lumber}} = \frac{\text{percent}}{100}$$

$$\frac{6\ inches}{8\ feet} = \frac{n}{100}$$

Since all the information must be in the same unit of measurement, do the conversion.

conversion:

$$\frac{1 \text{ foot}}{12 \text{ inches}} = \frac{8 \text{ feet}}{x}$$

$$x = 12 \times 8$$

$$x = \textbf{96 inches}$$

$$\frac{6 \text{ inches}}{96 \text{ inches}} = \frac{n}{100}$$

$$96n = 6 \times 100$$

$$96n = 600$$

STEP 4 Solve.

$$n = \frac{600}{96} = 6\frac{1}{4}\%$$

Note: In the conversion, the letter x was used to stand for the unknown number of inches. Any letter can be used to stand for an unknown.

Solve the following word problems, making the necessary conversions. Be careful; not all problems need a conversion. Refer to the conversion chart on page 221, if necessary.

1. Tile Town sells square-shaped 81-square-inch tiles. How many tiles are needed to cover a 54-square-foot floor?

2. On the airplane assembly line, Isadore was able to make 20 welds an hour. How many welds did he make during a 9-hour workday?

3. On Interstate Highway 501, there is a reflector every 528 feet. How many reflectors are there on the stretch of highway shown at the right?

   46 miles
   INTERSTATE HIGHWAY

4. A medical center needed 48 gallons of blood after an earthquake. A nearby city donated 26 gallons of blood. The rest was donated at the medical center by people each giving 1 pint of blood. How many people were needed to give a pint of blood at the center?

5. The Heat Coal Company distributed 38 tons of coal to the power plants of its customers in 1 day. It delivered 4,000 pounds of coal to each of the power plants. How many power plants received deliveries?

6. Sharon was able to type 463 numbers during a 5-minute timing for data entry. At this rate, how many numbers could she type in an hour?

7. Lynn brought 12 quarts of ice cream to the Fourth of July picnic. If she gives each person a 4-ounce serving of ice cream, how many people will get ice cream?

Solving Word Problems Containing Unnecessary Information

Unnecessary information is more difficult to spot in combination word problems than in one-step word problems. The key to identifying unnecessary numbers is in working backward from the question. Once you write a word sentence, look at all the given information, and decide what is needed to answer the question.

EXAMPLE An 0.8-ounce jar of basil sells for $0.98. Marie has 3.5 pounds of basil to pack into the jars. How many jars will she need?

STEP 1 *question:* How many jars will she need?

STEP 2 *necessary information:* 0.8 ounce, 3.5 pounds

(The cost of the jar of basil is unnecessary information.)

STEP 3 *solution statement:*

$$\frac{\text{total weight}}{\text{number of jars}} = \frac{\text{weight}}{1 \text{ jar}}$$

$$\frac{3.5 \text{ pounds}}{n \text{ jars}} = \frac{0.8 \text{ ounce}}{1 \text{ jar}}$$

Since all your weights must be in the same unit of measurement, your next step must be a conversion to find the number of ounces in a pound.

conversion: $\dfrac{16 \text{ ounces}}{1 \text{ pound}} = \dfrac{n \text{ ounces}}{3.5 \text{ pounds}}$

$$\frac{56 \text{ ounces}}{n \text{ jars}} = \frac{0.8 \text{ ounce}}{1 \text{ jar}}$$

$$1 \times n = 3.5 \times 16$$

$$0.8 \times n = 56 \times 1$$

$$n = 56 \text{ ounces}$$

$$0.8n = 56$$

STEP 4 Solve.

$$n = \frac{56}{0.8} = 70 \text{ jars}$$

Write the word sentences or proportion needed to solve the following word problems. Underline the necessary information. Then solve the problem. Be careful; many of these problems contain unnecessary information.

1. At the beginning of the school year, the Philadelphia school system had 103,912 students. During the course of the year, 4,657 students left the system, while 1,288 more students were enrolled. What was the student population at the end of the year?

2. At the beginning of the school year, the Philadelphia school system had 103,912 students. During the course of the year, 4,657 students left the system, while 1,288 more students were enrolled. How many different students spent at least part of the year in the Philadelphia school system?

3. At sunrise the temperature was 54 degrees. By midafternoon, it had risen 27 degrees. The temperature then began falling, until by midnight it had dropped 19 degrees from the high. What was the temperature at midnight?

4. Every week, after having $153 taken out of his paycheck, Lloyd takes home $548. What was Lloyd's take-home pay for a 52-week year?

5. Ahmed bought three paperbacks and two magazines at the drugstore at the prices shown at the right. He paid for his purchases with a $50 bill. How much change did he receive at the drugstore?

Hoyle's Drugstore	
Newspapers	$0.55
Paperbacks	$9.95
Magazines	$3.50
Postcards	$0.75

6. Sangita is a member of a cooperative grocery store. She gets a 20% discount off everything she buys in the store. She bought a 5-pound bag of oranges marked $4.80. After receiving her discount, how much did she pay for the oranges?

7. A one-cup serving of Cheerios weighs 28 grams and contains 20 grams of carbohydrates and 3 grams of dietary fiber. What percent of a serving of Cheerios is dietary fiber?

8. Juanita worked 40 hours at the Mexico Restaurant. She traveled an additional 5 hours to commute between work and home. She gets paid a salary of $6.40 an hour. She also earned $186 in tips for the week. How much did she earn for the week?

Solving Longer Combination Word Problems

Sometimes word problems cannot be solved by being broken into two one-step problems. Three or more steps may be needed to solve the problem. The method used with these problems is the same as the method used throughout this chapter with combination word problems. Keep working backward from the question. Set up a solution sentence and solve shorter problems to get all of the information that you need.

EXAMPLE Sylvia went shopping at the Bargain Basement. She bought a $54.99 dress marked $\frac{1}{3}$ off and a $36.95 pair of pants marked down 20%. How much did she spend?

STEP 1 *question:* How much did she spend?

STEP 2 *necessary information:* $54.99, $\frac{1}{3}$ off; $36.95, marked down 20%

STEP 3 *solution statement:*

dress price + pants price = total spent

To solve this, you must find the sale prices of both the dress and the pants. Both can be found by using this *missing information* sentence:

original price − discount amount = sale price

You can find the discount by multiplying the original amount by a fraction or percent.

dress

$54.99 − ($\frac{1}{3}$ × 54.99) = sale price

54.99 − 18.33 = $36.66

pants

$36.95 − (20% of 36.95) = sale price

36.95 − (0.20 × 36.95) = sale price

36.95 − 7.39 = $29.56

STEP 4 Solve.

$36.66 + $29.56 = **$66.22**

Note: An earlier chapter used the proportion method for solving percent word problems. However, if a problem requires you to find a percent of an amount, there is another method. Simply change the percent to a decimal and multiply. In the example above, 20% was changed to 0.20.

For every problem, write all necessary word sentences or proportions. Then solve the problem. Round money solutions to the nearest cent.

1. Kerry bought seven apples and a cantaloupe at the prices shown at the right. How much did she spend?

Cantaloupe	$2.88 each
Grapes	$2.69 per pound
Apples	$3.76 per dozen

2. Aaron received a bill of $96.80 for 32 gallons of bottled gas. If he pays the bill within 10 days, he will receive a 6% discount. How much will he pay if he pays his bill within 10 days?

3. Harold's doctor advised him to cut down on the calories he consumes by 28%. Harold has been consuming 4,200 calories a day. If Harold's breakfast contains 797 calories, how many calories can he have during the rest of the day?

4. Sarkis is a salesman. He receives a salary of $150 a week plus a 6% commission on all his sales over $400. Last week he sold $6,160 worth of merchandise. What was he paid for the week?

5. Dinora drove 3,627 miles from coast to coast. Her car averaged 31 miles per gallon, and she spent $338 for gasoline. On the average, what did she pay per gallon of gasoline?

Solving Combination Word Problems

In the following problems, choose the one best answer. Round decimals to the nearest hundredth.

1. Every day Pablo has to drive 7 miles each way to work and back. At work, he has to drive his truck on a 296-mile delivery route. How many miles does he drive during a 5-day workweek?

   a. 310 miles

   b. 315 miles

   c. 1,550 miles

   d. 4,214 miles

   e. none of the above

2. Every day, Jason has a 14-mile round-trip drive to work. He then has to drive his truck on a 296-mile delivery route 5 days a week. How many miles does he drive each day?

   a. 310 miles

   b. 315 miles

   c. 1,550 miles

   d. 4,214 miles

   e. none of the above

3. For his art class, Rajan spent $420 on books and $300 on materials. To cover costs, how much did each of his 15 students pay?

   a. $620

   b. $120

   c. $48

   d. $28

   e. $20

4. Jessie's restaurant had four small dining rooms with a seating capacity of 28 people each and a main dining room with a seating capacity of 94 people. What was the total capacity of the restaurant?

   a. 126 people

   b. 658 people

   c. 348 people

   d. 122 people

   e. 206 people

5. Each team in the 8-team football league used to have a roster of 36 players. The league decided to decrease each team's roster size by 3 players. Before the change, how many players were in the league?

   a. 180 players

   b. 288 players

   c. 285 players

   d. 264 players

   e. 396 players

6. Each team in the 8-team football league used to have a roster of 36 players. The league decided to decrease each team's roster size by 3 players. After the change, how many players were in the league?

   a. 180 players

   b. 288 players

   c. 285 players

   d. 264 players

   e. 396 players

7. At Meg's Market, Monique bought 2.36 pounds of cheese and 4 pounds of apples. What was the total cost of the cheese and apples at the prices shown at the right?

   a. $3.56

   b. $5.47

   c. $9.65

   d. $10.81

   e. $14.37

Meg's Market **On Sale This Week!**
Apples—only $0.89/pound Chicken—only $1.39/pound Potato Salad—only $1.29/pint Cheese—only $2.58/pound

8. A bottle contains 6 cups of laundry detergent. The directions say to use $\frac{1}{3}$ cup for a top-loading washer and $\frac{1}{4}$ cup for a front-loading washer. How many more loads per bottle can you do with a front-loading washer than with a top-loading washer?

   a. 1 load

   b. 3 loads

   c. 6 loads

   d. 8 loads

   e. 9 loads

Posttest A

This posttest gives you a chance to check your skills at solving word problems. Take your time and work each problem carefully. When you finish, check your answers and review any topics on which you need more work. (**Caution:** At least one problem does not contain enough information to solve the problem.)

1. Three tablespoons cocoa plus 1 tablespoon fat can be substituted for 1 ounce of chocolate in baking recipes. A recipe for chocolate cake calls for 12 ounces of chocolate. If Shirley is substituting cocoa for chocolate, how much cocoa should she use?

2. Matt has a 400-square-inch board. He needs a 25-square-inch piece of the board for the floor of a birdhouse. What percent of the board will he need for the floor of the birdhouse?

3. There are 5,372 school-age children in town. Of those children, 1,547 either go to private school, are homeschooled, or have dropped out. How many children remain in the town's public schools?

4. A bushel of apples weighs 48 pounds. Mariska wants to buy 12 pounds of apples. How many bushels should she buy?

5. At the sidewalk stand, Jason bought a hot dog and a soda. How much did he spend at the prices shown at the right?

HOT DOG $2.50
ITALIAN SAUSAGE $3.95
POTATO CHIPS $0.85
SODA $1.25

6. If Kenneth retires at age 65, he will receive as a pension 80% of his salary of $48,657. If he retires at age 62, he will receive only 70% of his salary. How much less will he receive for his pension if he retires early?

7. One cup sugar plus $\frac{1}{4}$ cup liquid can be substituted for 1 cup of corn syrup in baking recipes. A recipe calls for $1\frac{1}{2}$ cups corn syrup. If Mira is substituting sugar for corn syrup, how much liquid should she add?

8. A conservation organization charged each member $20 dues plus $15 for the organization's magazine. How much money did the organization collect from its 13,819 members?

9. Raul and Connie have $26,000 to use for a down payment on a new home. The bank told them their down payment must be at least 8% of the purchase price. What is the most expensive home they can afford?

10. Emily went on a shopping spree. She bought designer jeans for $134, a swimsuit for $98, and a wool sweater for $74. She then spent $13 at the food court. How much did she spend on clothes?

11. After 3 years, Anisa's car had lost $\frac{1}{3}$ of its original value. Two years later, it had lost an additional $\frac{1}{4}$ of its original value. If she bought the car for $12,900, what was its value after 5 years?

12. There are 3 feet in a yard. There are 1,760 yards in a mile. Kaliska is doing the 5-mile Walk for Peace. How many feet long is the Walk for Peace?

13. Tiva weighed $172\frac{1}{2}$ pounds. She lost $47\frac{3}{4}$ pounds in one year's time. What was her new weight?

14. To qualify for the car race, Christine needed to drive 100 miles in under 43 minutes. She completed the first lap in $4\frac{1}{2}$ minutes. At this rate, what will be her total time for the 100-mile qualifying distance?

15. Nickilena, Jean, Mei, and Elaine went into business together. The four-woman partnership earned \$336,460 and had expenses of \$123,188. If they divided the profits equally, how much did each person make?

16. Sears is offering 20% off on its \$768 refrigerator. How much can you save by buying the refrigerator on sale?

17. Before a recent election, $\frac{1}{3}$ of the voters polled said they were planning to vote for the incumbent, while $\frac{1}{4}$ said they were planning to vote for the challenger. The rest were undecided. What fraction of the voters had decided which way they were going to vote?

18. Denise is paid \$14.20 per hour at her part-time job. Last week she worked 17.5 hours. How much did she earn last week at her part-time job?

19. Janine bought a new car. She ordered $4,350 of added options and received a $2,650 rebate. How much did Jane pay for the car?

20. East Somerville has 948 homes. The Heart Association has 12 people collecting donations. If they all visit the same number of homes, how many homes should each of them visit?

21. Melvin received an electric bill for $116.29. He knows that it cost him $59.00 a month for his air conditioning. How much would his bill have been if he had not operated the air conditioner?

22. Eileen bought three pairs of socks for $3.79 each and four towels for $4.69 each. How much did she spend?

23. Enzo's Ice Cream Store puts $\frac{1}{16}$ pound of whipped cream on every sundae. For how many sundaes will the container of whipped cream pictured at the right last?

24. After paying $28.43 for dinner and $9.50 for a movie, Florence paid the babysitter $25.00. How much did the evening cost her?

25. A pile of books weighed 34.2 pounds. If each book weighed 0.6 pound, how many books were in the pile?

POSTTEST A CHART

Circle the number of any problem that you miss. A passing score is 20 correct answers. If you passed the test, go on to Using Number Power. If you did not pass the test, review the chapters in this book, referring to the practice pages listed below.

Problem Numbers	Skill Area	Practice Pages
1, 3, 10, 20	whole numbers	17–36, 57–71
4, 7, 13, 17, 23	fractions	48–56, 75–83
5, 18, 21, 24, 25	decimals	42–47, 52–56, 72–74, 82–83
2, 9, 16	percents	112–129
12	conversion	94–95, 148–149
14, 19	not enough information	104–108
6, 8, 11, 15, 22	multistep word problems	130–155

USING
NUMBER
POWER

Using Information from a Chart for Health

Even a relatively small chart can contain a great amount of useful information. Sometimes you will need to use your problem-solving and math skills to understand and utilize the information contained in a chart. The following chart has information that can have an impact on a person's health.

Chart of Beverages					
Serving Size	Beverage	Calories	Carbohydrates	Protein	Fat
8 fl oz	2% low-fat milk	113 cal	11 grams	7 grams	4 grams
12 fl oz	vanilla milk shake	381 cal	60 grams	13 grams	10 grams
8 fl oz	orange juice from concentrate	91 cal	22 grams	1 gram	0 gram
12 fl oz	cola	133 cal	34 grams	0 gram	0 gram

Georgiana wanted to gain weight. How many more calories would she get from a 12-fluid ounce serving of the highest calorie beverage compared to a 12-fluid ounce serving of the next highest calorie beverage?

At first glance, you might think that the cola is the next highest calorie beverage, but the serving size is larger than the milk or the orange juice. Since the milk and the orange juice are both 8-fluid ounce servings and the milk is higher in calories, you can eliminate the juice. You can now set up a proportion to find the calories in a 12-fluid ounce serving of low-fat milk.

$$\frac{8 \text{ fluid ounces}}{12 \text{ fluid ounces}} = \frac{113 \text{ calories}}{n \text{ calories}}$$

$$8n = 1,356$$

$$n = \textbf{169.5 calories} \text{ in 12 fluid ounces of milk}$$

So 12 fluid ounces of milk is the next highest calorie beverage.

Highest calorie beverage − next highest calorie beverage = difference in calories

$$381 \text{ calories} - 169.5 \text{ calories} = \textbf{211.5 calories}$$

Use the chart of beverages on page 162 to solve the following word problems.

1. For breakfast Danielle had 8 fluid ounces of orange juice. At lunch she had 12 fluid ounces of milk shake, and for a snack she drank 12 fluid ounces of cola. How many grams of carbohydrates did she consume from the three beverages?

    a. 32 grams

    b. 38 grams

    c. 116 grams

    d. 127 grams

    e. 505 grams

2. Keion had an 18-fluid ounce vanilla milk shake. How many grams of fat were in his milk shake?

    a. 10 grams

    b. 12 grams

    c. 13 grams

    d. 15 grams

    e. 16 grams

Use the nutrition chart below to solve questions 3–7.

	Calories	Carbohydrates	Protein	Fat
12-ounce sirloin steak	914	0 g	94 g	57 g
5-ounce hamburger on bun	405	24 g	27 g	21 g
6-ounce fried chicken breast	442	15 g	42 g	22 g
10-ounce spaghetti with tomato sauce	246	52 g	8 g	1 g
8-ounce broiled salmon	412	0 g	61 g	17 g

3. When Jean looked at the nutrition chart, she had trouble comparing the different main dishes because the portions were different sizes. She was trying to decide whether to make a 12-ounce portion of sirloin steak or a 12-ounce portion of broiled salmon. How many more calories was the steak than the salmon?

    a. 914 calories

    b. 1,326 calories

    c. 502 calories

    d. 296 calories

    e. 1,532 calories

4. Elba wanted to prepare a main dish that was approximately 600 calories. If she prepared one of the main dishes from the chart, how large should a portion of that main dish be?

5. Max was going to have a 10-ounce portion of sirloin steak, but then decided he needed to cut down on the amount of fat in his diet. How much less fat would he eat if he had a 10-ounce portion of one of the other main dishes on the chart?

6. Using the chart as a guide, where do the grams of carbohydrates in a hamburger on bun come from, the hamburger or the bun or both?

7. Anya decided she needed more protein in her diet. She also wanted to have less fat in her diet. Which entrée would be the best choice for her?

   a. sirloin steak

   b. hamburger on bun

   c. fried chicken breast

   d. spaghetti with tomato sauce

   e. broiled salmon

Use the exercise chart below to answer questions 8–12.

Exercise for a 130-Pound Woman	
Walking 3 mph	5 calories/minute
Jogging 5.5 mph	10 calories/minute
Running 10 mph	19 calories/minute
Swimming 25 yd/min	4 calories/minute
Bicycling 6 mph	4 calories/minute
Tennis singles	6 calories/minute
Cross-country skiing	6 calories/minute
Aerobic dancing	8 calories/minute
Recreational volleyball	3 calories/minute
Sleeping	1 calorie/minute

Match questions 8 through 10 to the correct solutions listed below.
Some questions have more than one solution.

> **a.** recreational volleyball for 90 minutes + tennis for 55 minutes
>
> **b.** running for 25 minutes + walking for 45 minutes + bicycling for 25 minutes
>
> **c.** jogging for 40 minutes + swimming for 30 minutes + bicycling for 2 hours
>
> **d.** aerobic dancing for 40 minutes + tennis for 40 minutes + jogging for 24 minutes
>
> **e.** cross-country skiing for 2 hours + aerobic dancing for 35 minutes

8. How could Fernande burn 1,000 calories? _____

9. Rafaela wanted to spend no more than 2 hours doing at least three different activities burning a total of 800 calories. How could she do it? _____

10. Juliette wanted to burn 600 calories while exercising at least 2 hours. How could she do it? _____

11. Design a fitness program for yourself using the numbers from the table (even if you are not a 130-pound woman). Plan to burn exactly 500 calories in at least an hour of activity.

12. Design an exercise program that is exactly one hour long and consists of at least two different activities. How many calories would a 130-pound woman burn if she followed your exercise program?

Using Number Power in a Landscaping Business

You need number power in almost every business. Landscaping is a business that uses mathematics in many different ways. Frank's Landscaping and Lawn Care handles a variety of jobs, including putting up fences; planting trees, shrubs, and flowerbeds; seeding and fertilizing lawns; and troubleshooting lawn and garden care problems. Frank also has to manage the financial side of his business, including controlling costs and pricing his jobs so that he makes enough money to support his family and to pay his workers.

Frank's customer, Yao, needs to have a five-foot-high fence put around his 26-foot by 18-foot vegetable garden to keep out deer. How many feet of fencing does Frank need for the job?

In order to find out how much fencing Frank needs for the job, you have to calculate the distance around, or perimeter of, the garden. If he walked around the boundary of the garden, the distance would be 26 feet + 18 feet + 26 feet + 18 feet, or 88 feet. The longer distance is called the length, and the shorter distance is called the width. Notice that each distance appears twice. The perimeter, P, is equal to 2 lengths plus 2 widths, or $P = 2l + 2w$.

Frank recommends one pound of grass seed for every 200 square feet when seeding a new lawn. He needs to seed a new lawn on a rectangular 40-foot by 90-foot front yard. How many pounds of grass seed should he use?

Solution sentence: $\dfrac{1 \text{ pound grass seed}}{200 \text{ square feet of lawn}} = \dfrac{n \text{ pounds}}{\text{area of lawn}}$

To find the area of a rectangular lawn, multiply the width $\times$ the length.

40 feet $\times$ 90 feet = **3,600 square feet**

$$\frac{1 \text{ pound}}{200 \text{ square feet}} = \frac{n \text{ pounds}}{3,600 \text{ square feet}}$$

Solve: $200n = 3,600$

                     $n =$ **18 pounds**

For each of the following word problems, write a solution sentence and solve.

1. Frank is putting fertilizer on a customer's lawn that is 35-feet by 75-feet. What is the area of the lawn?

2. Frank uses 40 pounds of organic fertilizer for every 1,000 square feet of lawn. How many pounds of fertilizer does he need for Larry and Cali's 30-foot by 80-foot lawn?

3. Frank charges $35 an hour for labor. He worked for 5 hours putting fertilizer on the lawn and also charged at his labor rate for $\frac{1}{2}$ hour of travel. How much did Frank charge for labor?

4. Frank buys 40-pound bags of fertilizer wholesale for $28 a bag. For fertilizing jobs, he marks up the price of a bag 60%. How much does Frank charge his customers for each bag of fertilizer?

5. Frank's customer, Domingo, wants to have a fence put up in his backyard so that his dog has a place to run. The backyard is a 50-foot by 120-foot rectangle as pictured in the diagram. The house is 100 feet long. How many feet of fencing are needed to enclose the backyard? (**Note:** Frank does not have to put fencing along the back of the house.)

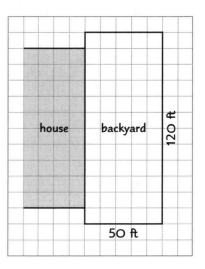

6. Frank charges Danielle $2.60 per linear foot for chain-link fencing. How much should Frank charge Danielle for 316 linear feet of chain-link fencing?

Questions 7 and 8 are based on the chart at the right.

7. On one landscaping job, Frank paid Kendrick and Marquis for 7 hours of work. How much more did Kendrick earn than Marquis?

Frank's Workers and Their Rate of Pay	
Shakira	$15.40 an hour
Kendrick	$11.62 an hour
Marquis	$9.78 an hour

8. Shakira handles Frank's office work, including doing all of the bookkeeping. Last week she worked 32.5 hours. How much did she earn last week?

Using Number Power in Retail

Number power is very important in running a retail business. Products need to be bought at wholesale prices. Markups need to be determined so that the business is profitable and customers buy the products. Inventories must be maintained. Rent, utilities, salaries, and commissions must be paid.

Sia manages the High Fashion/Low Prices clothing store. She purchased 30 shirt-and-shorts outfits for $750. How much did each outfit cost?

Solution sentence:

$$\frac{\text{total price of outfits}}{\text{number of outfits}} = \text{price of one outfit}$$

$$\frac{\$750}{30 \text{ outfits}} = \text{price of one outfit}$$

$$\mathbf{\$25} = \text{price of one outfit}$$

For each of the following word problems, write a solution sentence or proportion and then solve.

1. Sia paid $35 each for skirt-and shirt-outfits. She marked up the price by 50%. How much did she charge for each outfit?

2. Sia ran a sale in which everything in the store was 20% off. Designer jeans were marked $84. If Jennifer bought a pair of these designer jeans, how much would she save from the marked-up price?

3. At the end of the season, Sia ran a 65% off Clearance Sale. A bathing suit was originally priced at $96. What was its clearance price?

4. Last month, Sia spent $892.80 on rent and utilities, $5,193.85 on salaries, and $3,645 on inventory and other expenses. How much does she need to earn in sales to break even for the month?

5. Ramona bought a jacket for $68.99 and a pair of gloves for $22.49. She gave Sia a $100-bill to pay for her purchases. How much change should Sia give Ramona?

In addition to its clothing store, High Fashion/Low Prices also sells its merchandise on its Web site. Purchases include shipping charges summarized on the following chart:

SHIPPING CHARGE	
Total Order Amount	**Standard Shipping**
up to $14.99	$6.95
$15 to $37.99	$8.95
$38 to $74.99	$12.95
$75 to $99.99	$14.95
$100 to $149.99	$19.95
$150 to $199.99	$27.95
$200 to $299.99	$39.95
$300 to $399.99	$54.95
$400 and up	$69.95

6. Alisa bought a scarf for $12.95, a belt for $17.49, and a vest for $36.89. Including shipping, how much did she have to pay for her purchase?

7. Laurie bought a suit for $119.65, 3 tops for $27.85 each, and 4 pairs of crew socks for $3.49 each. Including shipping, how much did she have to pay for her purchase?

8. High Fashion/Low Prices allows customers to pay for purchases through the PayFriend service. The PayFriend service charges High Fashion/Low Prices 2% of the total amount paid. Alisa and Laurie (Questions 6 and 7) both used PayFriend to pay for their online purchases. How much did High Fashion/Low Prices have to pay PayFriend for Alisa's purchase (Question 6)?

9. How much did High Fashion/Low Prices have to pay PayFriend for Laurie's purchases (Question 7)?

Using Number Power in the Medical Field

Number power is used at medical care facilities every day. Doctors and nurses need to be able to calculate correct dosages of medications for patients. Medical workers need to be able to read a variety of monitors and interpret the numerical outputs.

EXAMPLE Jean works as a nurse at the Total Care Nursing Home. She has to give medication to patients based on the written orders from doctors. Medication doses are usually in milligrams, or mg. The active ingredient is usually dissolved in a solution that is measured in milliliters.

The doctor ordered that Doris get a 0.5 mg dose of Ativan to treat anxiety. In the nursing home, Ativan comes in a vial that contains 2 mg of Ativan in 1 ml of solution. How many milliliters of solution should Jean draw up in the syringe to inject in Doris?

Set up a proportion: $\dfrac{2 \text{ mg Ativan}}{1 \text{ ml solution}} = \dfrac{0.5 \text{ mg}}{n \text{ ml solution}}$

Cross multiply, then divide: $2n = 0.5$

$n = \textbf{0.25 ml solution}$

For each of the following word problems, write a solution sentence or proportion and solve.

1. The doctor ordered 650 mg of Tylenol to be given every 4 hours for 24 hours to treat Stella for pain and fever. Jean has 325 mg Tylenol tablets. How many tablets will she need for Stella for one day?

2. The doctor ordered 2 liters of intravenous fluids to be given to Arnold over a 24-hour period. How many milliliters of fluid should Arnold receive per hour?

3. The doctor orders 0.125 mg of Digoxin for Margaret to be taken by mouth once a day in order to control congestive heart failure. Jean has 0.25 mg tablets of Digoxin. How many tablets should Jean give Margaret each day?

4. Lillian has diabetes and has to monitor her blood sugar 4 times a day. She uses 1 test strip for each monitoring. How many days would a bottle of 50 strips last?

5. When Jean worked in the acute care hospital, the doctors always asked for a patient's temperature in degrees Celsius. The patient would usually want to know the temperature in Fahrenheit. The conversion formula for Celsius to Fahrenheit is $°F = \frac{9}{5}°C + 32$. Italio had a temperature of 39° Celsius. What was his temperature in Fahrenheit?

6. Jean's friend Patricia is a Nurse Practitioner in a pediatric clinic. She prescribes medication doses based on the weight of the child she is treating. The chart she uses gives weight ranges in kilograms, but the scale in the office is in pounds. Mona weighs 48 pounds. How many kilograms does she weigh?

7. Patricia prescribes amoxicillin to treat Mona's ear infection. The daily dosage needs to be 20 mg per kilogram that the patient weighs. What daily dosage should Patricia prescribe for Mona?

8. Lamine has a severe cough. His doctor prescribed a codeine cough medicine. The bottle label says that each 5 ml contains 10 mg of codeine. Since Lamine is 2 years old, his dosage should be 5 mg of codeine every 4 hours. How many milliliters of cough medicine should he be given every 4 hours?

9. The doctor ordered 0.6 mg of Atropine for Eduardo. Jean has on hand Atropine 2 mg/ml. How many milliliters of Atropine should Jean give Eduardo?

10. Jean mixed 1.8 ml of saline with 1 g of Oxacillin powder to create 0.5 g to 1 ml liquid Oxacillin solution. She needed to give her patient, Angel, 750 mg of Oxacillin. How many milliliters of the solution should Jean give Angel?

Using Number Power in the Trucking Industry

You need number power in the trucking industry. You need to estimate the costs so that you know what you need to charge your customers. If you have a fleet of trucks, you need number power to figure out how to use them most efficiently.

Sal's Long and Short Haul Trucking has a fleet of trucks that make deliveries throughout the Northeast. Their Class 8 trucks, on average, are driven 120,000 miles per year and get 6 mpg. If fuel costs $4.15 per gallon, on average, how much does fuel cost for one Class 8 truck per year?

Question: How much does fuel cost for one Class 8 truck per year?

Solution sentence: number of gallons × cost per gallon = total cost for fuel

$$\frac{\text{miles driven}}{\text{mileage}} = \text{number of gallons}$$

$$\frac{120,000 \text{ miles}}{6 \text{ mpg}} = \text{number of gallons}$$

$$20,000 = \text{number of gallons}$$

20,000 gallons × $4.15 per gallon = total cost for fuel

$83,000 = total cost of fuel

For each of the following word problems, write a solution sentence or proportion and solve.

1. The Class 8 truck, which gets 6 mpg, is hauling a load of high definition televisions 1,410 miles. How many gallons of fuel will the truck burn for this distance?

2. For one delivery, Sal's Trucking charged a flat $200 fee plus $5 per mile. The truck had to travel 52 miles. How much did Sal's charge for the delivery?

3. Sal's owns a Volvo truck that has a 600-liter fuel tank. How many gallons of gas would be in a filled fuel tank? (**Hint:** 1 liter = 0.26 gallon)

4. Sal's Volvo truck has a speedometer that gives a speed reading in kilometers per hour. The speed limit on an interstate is 65 miles per hour. What is the speed limit in kilometers per hour? (**Hint:** 1 mi = 1.61 km)

Questions 5–7 are based on the map below.

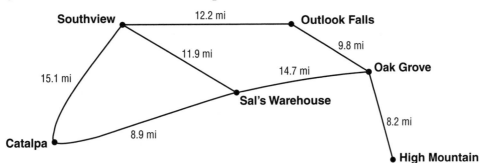

5. Starting from Sal's Warehouse, driver Melissa needs to make deliveries to Catalpa and Southview and then return to the warehouse. Which is her shortest delivery route?

6. Starting from Sal's Warehouse, driver Felipe needs to make deliveries to Oak Grove, High Mountain, Outlook Falls, and Catalpa, and then return to the warehouse. How many miles is the shortest delivery route?

7. Starting from Sal's Warehouse, driver Francisco needs to pick up merchandise from Outlook Falls and bring it back to the warehouse. How many miles is his shortest round trip?

Challenging Word Problems

Sometimes word problems require more than translating words into mathematics. You might have to do more complex problem solving.

EXAMPLE Juana's recipe for chili calls for 2 teaspoons of hot sauce per gallon. Unfortunately, she misread the instructions and put in 2 tablespoons of hot sauce. How much more chili should she make in order to readjust her chili back to the recipe?

STEP 1 *question:* How much more chili should she make in order to readjust her chili back to the recipe?

STEP 2 *necessary information:* 2 teaspoons per gallon, 2 tablespoons

STEP 3 Decide what arithmetic operation to use. Convert tablespoons to teaspoons. Then use a proportion to find the total amount of chili.

final equation: total amount of chili − chili already made = additional chili

STEP 4 There are 3 teaspoons in a tablespoon.

$$\frac{3 \text{ teaspoons}}{1 \text{ tablespoon}} = \frac{n \text{ teaspoons}}{2 \text{ tablespoons}}$$

$$n = 6 \text{ teaspoons}$$

Calculate the new total amount of chili.

$$\frac{2 \text{ teaspoons}}{1 \text{ gallon}} = \frac{6 \text{ teaspoons}}{x \text{ gallons}}$$

$$2x = 6$$

$$x = 3 \text{ gallons}$$

Calculate how much more chili needs to be made.

1 gallon chili + g gallons to be made = 3 gallons of chili

$$g = 2 \text{ gallons of chili}$$

STEP 5 Does the answer make sense?

If three times the original amount of hot sauce is added to the recipe, you should end up with three times the original amount of chili. You will have correctly readjusted the recipe.

Solve the following word problems.

1. After a large storm, Pablo had 8 inches of water in his 30-foot by 40-foot basement. His sump pump pumped out 200 cubic feet of water an hour. Assuming no more water seeped into the basement, how long did it take to pump all the water out of the basement?

2. The same storm left 9 inches of water in Chantel's 30-foot by 50-foot basement, and some water continued to seep in for 4 hours. Her sump pump pumped out 300 cubic feet of water an hour. It took 6 hours to pump the water out of her basement. How much water seeped into the basement?

3. Awilda needs to get gasoline for her car, buy groceries, shop for a blouse at the department store, take out a book from the library, and return a video to the video store. According to the map, what is the shortest total distance she will have to drive to do all her chores and return home?

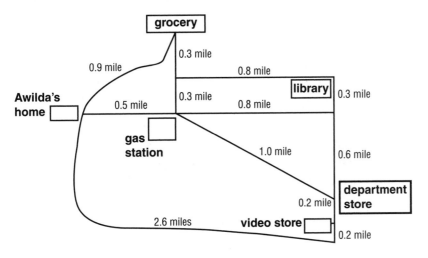

4. It takes 5 seconds for the sound of thunder to travel one mile. Light travels so fast that we can assume we are seeing lightning instantly. A storm is rapidly approaching. Domingo sees a lightning flash and hears the thunder 15 seconds later. Ten minutes later he sees another lightning flash and hears thunder 10 seconds later. If the storm continues to move at the same rate, how long will it take before it is directly overhead?

5. John is designing a 400-square-foot rectangular garden. He wants the garden to have the smallest possible perimeter, so he can spend as little as possible on fencing. What should the dimensions of his garden be?

6. Carmen has 100 feet of fencing for a dog run. The run has to be at least 3 feet wide. How long could the run be?

7. Pitcher Jonathan Papelbon can throw a 95 miles-per-hour fastball. It is 60.5 feet from the rubber at the top of the pitcher's mound to home plate. How long does it take Papelbon's fastball to travel from the pitcher's mound to home plate?

Posttest B

This test has a multiple-choice format much like the GED and other standardized tests. Take your time and work each problem carefully. Round decimals to the nearest hundredth. Circle the correct answer to each problem. When you finish, check your answers at the back of the book.

1. Large eggs weigh $1\frac{1}{2}$ pounds per dozen. Gia bought eight large eggs. How much did the eggs weigh?

   a. 3 ounces

   b. $\frac{1}{3}$ pound

   c. 1 pound

   d. 18 ounces

   e. not enough information given

2. Glenn, the owner of a hardware store, originally paid $540.60 for 15 socket sets. At his year-end clearance sale, he sold the last socket set for $24.00. How much money did he lose on the last tool set?

   a. $180.60

   b. $1.50

   c. $12.04

   d. $36.04

   e. none of the above

3. Manny was working as a hot dog vendor. He sold a total of 426 hot dogs in one weekend. If he sold 198 on Saturday, how many did he sell on Sunday?

   a. 624 hot dogs

   b. 332 hot dogs

   c. 228 hot dogs

   d. 514 hot dogs

   e. none of the above

4. The television announcer reported that Elizabeth Quezada had received 39% of the votes in the race for mayor. The totals board behind the announcer showed that Elizabeth had received 156,000 votes. How many votes were cast in the election?

   a. 40,000 votes

   b. 400,000 votes

   c. 156,039 votes

   d. 608,480 votes

   e. 60,840 votes

5. On average, Morriston Airport has 96 jumbo jets arriving each day. Each jumbo jet has an average of 214 passengers. How many passengers arrive by jumbo jet at Morriston Airport each day?

    a. 310 passengers

    b. 20,544 passengers

    c. 118 passengers

    d. 222 passengers

    e. none of the above

6. To finish off the room, Ed needs a tile only $\frac{1}{3}$-foot wide. How much does he have to cut off the tile pictured at the right so that it will fit?

    a. $\frac{1}{2}$ foot

    b. $\frac{5}{12}$ foot

    c. $\frac{4}{7}$ foot

    d. $\frac{1}{4}$ foot

    e. $\frac{4}{9}$ foot

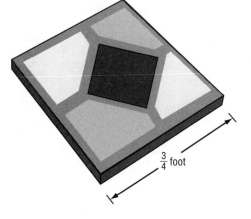

$\frac{3}{4}$ foot

7. After picking a bushel of apples, Tina plans to divide the apples equally among herself and five of her neighbors. How many apples will each of them get if the bushel weighs 54 pounds?

    a. 9 apples

    b. 60 apples

    c. 48 apples

    d. 324 apples

    e. not enough information given

8. Sergio uses 1.23 cubic yards of concrete to cover 100 square feet with 4 inches of concrete. How many cubic yards does he need to cover 550 square feet with 4 inches of concrete?

    a. 650 square feet

    b. 4.92 cubic yards

    c. 6.77 cubic yards

    d. 2,200 square inches

    e. 27.06 cubic yards

9. Pedro needs 4 pounds of hamburger for his chili recipe. In his freezer, he has a 2.64-pound package of hamburger. How much more hamburger does he need?

   a. 6.64 pounds

   b. 2.68 pounds

   c. 1.36 pounds

   d. 1.52 pounds

   e. 10.56 pounds

10. Last year the city's supermarkets sold 1,638,000 gallons of milk. There are 78,000 people in the city. On the average, how many gallons of milk did each person buy?

    a. 1,716,000 gallons

    b. 1,560,000 gallons

    c. 21 gallons

    d. 127,716 million gallons

    e. not enough information given

11. After having $198.23 deducted from his paycheck, Maurice takes home $532.77 every week. What are Maurice's total gross earnings for a 52-week year?

    a. $731.00

    b. $10,307.96

    c. $17,396.08

    d. $27,704.04

    e. $38,012.00

12. In the last year 423 service stations in the state closed. Only 2,135 remain. How many service stations existed in the state a year ago?

    a. 2,558 service stations

    b. 1,712 service stations

    c. 2,312 service stations

    d. 2,512 service stations

    e. none of the above

13. A two-thirds majority of those voting in the House of Representatives is needed to override a presidential veto. If all 435 representatives vote, how many votes are needed to override a veto?

    a. 290 votes

    b. 145 votes

    c. 657 votes

    d. 658 votes

    e. 224 votes

14. Naomi had $61 in her checking account. She wrote a check for $28 and made a deposit. How much money did she then have in the account?

    a. $89

    b. $33

    c. $2.18

    d. $1,708

    e. not enough information given

15. Avi needed to replace the molding on the left side of his car after it was damaged. The door needed $33\frac{3}{4}$ inches of molding, while the rear quarter panel needed $51\frac{2}{3}$ inches. How many inches of molding did he need if he replaced the molding on the door and the rear quarter panel?

    a. $2{,}247\frac{5}{12}$ inches

    b. $85\frac{5}{12}$ inches

    c. $17\frac{11}{12}$ inches

    d. $\frac{87}{124}$ inch

    e. $\frac{124}{87}$ inches

16. Ben, who works at the meat counter at the local supermarket, had to price the meat yesterday because the machine that normally did the job was broken. What price should he put on a 2.64-pound rib roast selling at $3.96 a pound?

    a. $1.50

    b. $5.94

    c. $6.60

    d. $10.45

    e. $1.32

17. At the factory, 28% of the workers were women. There were 432 male workers. What was the total number of workers at the factory?

    a. 460 workers

    b. 12,096 workers

    c. 600 workers

    d. 1,543 workers

    e. 404 workers

18. Sarah bought the carton of nails pictured at the right. How much did each nail weigh?

    a. 56 pounds

    b. 100 pounds

    c. 0.01 pound

    d. $\frac{1}{56}$ pound

    e. $\frac{3}{100}$ pound

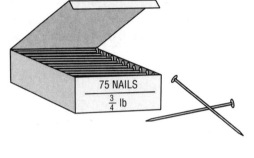

75 NAILS

$\frac{3}{4}$ lb

19. In the first quarter, the Philadelphia 76ers hit only 7 out of 25 field goal attempts. What was their scoring percentage?

    a. 28%

    b. 32%

    c. 72%

    d. 76%

    e. 18%

20. Carla gained 3 pounds in the first month of her new diet and 4 pounds in the second month. Her original weight was 104 pounds. What was her new weight?

    a. 97 pounds

    b. 105 pounds

    c. 103 pounds

    d. 111 pounds

    e. 100 pounds

1.23 pound steak

21. Peg bought the roast and the steak shown at the right. How much meat did she buy?

    a. 4.92 pounds

    b. 3 pounds

    c. 4.54 pounds

    d. 2.46 pounds

    e. 0.33 pound

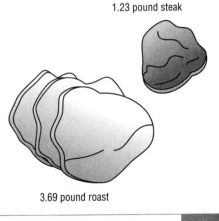

3.69 pound roast

22. Tamara had $74.81 in her checking account. She wrote checks for $46.19 and $22.45. She then made a $60.00 deposit. What was her new balance?

    a. $203.45

    b. $66.17

    c. $83.45

    d. $53.83

    e. $38.55

23. Premium Ice Cream is 9% milkfat. How many pounds of milkfat are in a 450-pound batch of Premium Ice Cream?

    a. 50 pounds

    b. 441 pounds

    c. 459 pounds

    d. 40.5 pounds

    e. 2 pounds

24. At the start of a trip, Tony filled his gasoline tank. After driving 168 miles, he needed 5.6 gallons of gasoline to refill his tank. How many gallons of gasoline would he use for the 417-mile drive from his home to his brother's home?

    a. 44.5 gallons

    b. 13.9 gallons

    c. 74.5 gallons

    d. 8.3 gallons

    e. 19.5 gallons

25. A piece of cheese was labeled $4.79 a pound. The price of the cheese was $2.06. How much did the cheese weigh?

    a. $9.87

    b. $6.85

    c. $2.73

    d. 2.33 pounds

    e. 0.43 pound

POSTTEST B CHART

If you missed more than one problem on any group below, review the practice pages for those problems. Then redo the problems. If you had a passing score, redo any problem you missed.

Problem Numbers	Skill Area	Practice Pages
3, 12, 20	add or subtract whole numbers	17–36
5, 10	multiply or divide whole numbers	57–71
6, 15	add or subtract fractions	37–41, 48–56
13, 18	multiply or divide fractions	75–83
9, 21	add or subtract decimals	37–47, 52–56
16, 25	multiply or divide decimals	72–74, 82–83
4, 19, 23	percents	112–129
1	conversion	94–95, 148–149
7, 14	not enough information	104–108
2, 8, 11, 17, 22, 24	multistep word problems	130–155

Pages 1–7, Pretest

1. b. weight per subcompact × number of
subcompacts = total weight
$1,600 \times 840 =$ **1,344,000 pounds**

2. b. current height − growth = height at
beginning of year
$48\frac{3}{8} - 2\frac{1}{4} = \mathbf{46\frac{1}{8}}$ **inches**

3. e. total cost per person × number of friends =
total collected
cost to get in + cost of inline skates =
cost per person
$7.50 + 5.00 = 12.50$
$\mathbf{12.50 \times 19 = \$237.50}$

4. a. $\dfrac{\text{part}}{\text{whole}} = \dfrac{\text{percent}}{100}$

$\dfrac{\$79.50}{\$n} = \dfrac{75}{100}$

$75 \times n = 79.50 \times 100$
$n = 7,950 \div 75 =$ **$106**

5. d. $\dfrac{\text{concentrate}}{\text{water}} = \dfrac{\text{concentrate}}{\text{water}}$

$\dfrac{5 \text{ tablespoons}}{2 \text{ cups}} = \dfrac{n \text{ tablespoons}}{24 \text{ cups}}$

$2 \times n = 5 \times 24$
$n = 120 \div 2 =$ **60 tablespoons**

6. c. $\dfrac{\text{part}}{\text{whole}} = \dfrac{\text{percent}}{100}$

$\dfrac{\$684}{\$3,600} = \dfrac{n}{100}$

$3,600 \times n = 684 \times 100$
$n = 68,400 \div 3,600 =$ **19%**

7. c. turkey + chicken = cost
$10.34 + 5.17 =$ **$15.51**

8. e. $\dfrac{200,000 \text{ pellets}}{n \text{ boxes}} = \dfrac{400 \text{ pellets}}{1 \text{ box}}$

$400 \times n = 1 \times 200,000$
$n = 200,000 \div 400 =$ **500 boxes**

9. d. *conversion:* 1 dozen oranges = 12 oranges
$\dfrac{\$2.40}{12 \text{ oranges}} = \dfrac{\$n}{4 \text{ oranges}}$

$12 \times n = 2.40 \times 4$
$n = 9.60 \div 12 =$ **$0.80**

10. d. no. of pounds ÷ no. of slices = wt of slice
$3.15 \text{ pounds} \div 24 \text{ slices} =$ **0.13 pound**

11. e. **not enough information given.** You need
to know how many inches are on a roll of
masking tape.

12. a. girder weight × number of girders =
total weight
$\dfrac{7}{8} \times 600 =$ **525 tons**

13. b. sale price + reduction = original price
$190 + 98 =$ **$288**

14. d. $\dfrac{n \text{ undecided}}{\text{total people}} = \dfrac{\text{percent undecided}}{100}$

$100\% - (\text{in favor} + \text{against}) =$
percent undecided
$100\% - (68\% + 25\%) = 100\% - 93\% =$
7% undecided

$\dfrac{n}{1,400 \text{ people}} = \dfrac{7}{100}$

$100 \times n = 1,400 \times 7$
$n = 9,800 \div 100 =$ **98 people**

15. b. $\dfrac{1\frac{1}{4} \text{ pounds}}{\$7.80} = \dfrac{1 \text{ pound}}{\$n}$

$\dfrac{5}{4} \times n = 7.80 \times 1$
$n = 7.80 \times \dfrac{4}{5} =$ **$6.24**

16. c. tank size − new gasoline = old gasoline
$12 - 6.78 =$ **5.22 gallons**

17. b. original price − reduction = sale price
fraction off × original price = reduction
$\dfrac{1}{3} \times 96 = 32$
$96 - 32 =$ **$64**

18. a. current rent − increase = old rent
$780 - 65 =$ **$715**

19. a. total tablets − tablets taken = tablets left
tablets/day × no. of days = tablets taken
$4 \times 30 = 120$ tablets taken
$250 - 120 =$ **130 tablets left**

20. c. *conversion:* 16 ounces = 1 pound
$\dfrac{2 \text{ ounces}}{1 \text{ box}} = \dfrac{n \text{ ounces}}{1000 \text{ boxes}}$

$1 \times n = 1,000 \times 2$
$n = 2,000$ ounces
$2,000 \div 16 =$ **125 pounds**

21. a.
$$\frac{\text{part}}{\text{whole}} = \frac{\text{percent}}{100}$$
$$\frac{n \text{ questionnaires}}{6{,}000 \text{ questionnaires}} = \frac{15}{100}$$
$$100 \times n = 15 \times 6{,}000$$
$$n = 90{,}000 \div 100$$
$$n = \textbf{900 questionnaires}$$

22. d. original gallons − gallons delivered = gallons left
deliveries × gallons per delivery = gallons delivered
$7 \times 364 = 2{,}548$ gallons delivered
$9{,}008 - 2{,}548 = \textbf{6,460 gallons left}$

23. c. 1st dress + 2nd dress = total fabric
$2\frac{5}{8} + 2\frac{3}{4} = \textbf{5}\frac{\textbf{3}}{\textbf{8}} \textbf{ yards}$

24. e. no. of pounds × price/pound = total cost
$1.62 \times 2.43 = \textbf{\$3.94}$

25. b. total aid ÷ no. of students = aid/student
last year's aid − decrease = total aid
$1{,}126{,}200 - 462{,}000 = 664{,}200$ total aid
$664{,}200 \div 820 = \textbf{\$810.00}$

26. c. original price − savings = sale price
Since the item is over 40 days old, the discount is 75%.
$$\frac{\text{savings}}{\text{original price}} = \frac{\text{discount}}{100}$$
$$\frac{s}{\$18} = \frac{75}{100}$$
$$100s = 18 \times 75$$
$$s = 1{,}350 \div 100 = \$13.50$$
$$18 - 13.50 = \textbf{\$4.50}$$

27. e. original price − savings = sale price
red suit: $\$40 - (0.25 \times \$40) = \$30$
floral print suit: $\$30 - (0.10 \times \$30) = \$27$
violet suit: $\$45 - (0.40 \times \$45) = \$27$
striped suit: $\$28 - (0 \times \$28) = \$28$
black suit: $\$60 - (0.75 \times \$60) = \textbf{\$15}$

28. e. not enough information given. You don't know the price of the sweater.

29. c. total bill = customer charge + kWh (delivery charge + supplier charge)   total bill =
$\textbf{\$5.81 + 445 kWh (\$0.047 + \$0.032)}$

30. c. $\dfrac{\text{total cost}}{\text{cost per movie}}$ = total movies
$\frac{8.99}{2.99} = 3$ movies
3 movies × 2.99 per movie = $8.97, which is a slightly better deal than $8.99.

Pages 12–13

1. How much snow fell during the entire winter?

2. What is the total cooking time?

3. Find the cost of parking at the meter for 3 hours.

4. How many years did Joe serve in prison?

Answers may vary. Sample answers:

5. Find the number of flats she can buy.

6. How much did the city receive for the delivery?

7. How many coats can be made from a roll?

8. How many grams of acetominophen are in a normal adult dose?

9. What was the price reduction?

10. How many yards of fabric can be produced in an hour?

11. How many more cars are sold during the Presidents' Week sale?

12. How much would you have to pay for a super pretzel and a soda?

Page 14

1. numbers: 124, 119; labels: commuters; answer label: commuters

2. numbers: 14, 5; labels: potatoes, lb; answer label: pounds (or lb)

3. numbers: 60, 250; labels: pages; answer label: pages

Page 16

1. *given information:* 22 years, 20 years, 23 years
necessary information: 22 years, 20 years. You are only comparing Mona's age with her sister's age. Therefore, the boyfriend's age is not needed.

2. *given information:* $386, $167, 2 children
necessary information: $386, $167. You do not need to know Rena's number of children in order to find the total amount of assistance she receives.

3. *given information:* 3 times, 20-year old, 10 hours
necessary information: 3 times, 10 hours. You do not need to use Laura's age to find out how many hours Marilyn works.

4. *given information:* 7-year old, $143, $139, $140, $131, first 2 months; *necessary information:* $143, $139. The chart shows information for 4 months; you are asked for information about the first months—January and February. The age of the car is not necessary.

5. *given information:* $2,400, $135,800, $3.60 per gallon; *necessary information:* $2,400, $3.60 per gallon. The money made on shoes is not needed to compute the gallons of oil bought.

6. *given information:* 45 years old, 8 people, half the family; *necessary information:* 8 people, half the family. Gia's age is not needed to answer the question.

7. *given information:* 4,700 workers, 3,900 skilled laborers, 700 people; *necessary information:* 4,700 workers, 700 people. To find the number of employees currently working, the total number of skilled workers is not necessary information.

8. *given information:* $2.49, 12-ounce glasses, 64 fl oz; *necessary information:* 12-ounce, 64 fl oz. The cost of the cola is not needed to find the number of glasses Maritza can fill.

Page 18

1. plus
2. and, altogether
3. more, altogether
4. increase
5. added, extra
6. in all

Page 19

1. key word: altogether
$$\begin{array}{r} 5 \text{ inches} \\ + 23 \text{ inches} \\ \hline \mathbf{28 \text{ inches}} \end{array}$$

2. key word: total
$$\begin{array}{r} 4,500 \text{ pounds} \\ + 836 \text{ pounds} \\ \hline \mathbf{5,336 \text{ pounds}} \end{array}$$

3. key word: and
$$\begin{array}{r} \$529 \\ + \$449 \\ \hline \mathbf{\$978} \end{array}$$

Page 20

1. key words: cheaper than
2. key word: decrease
3. key words: less than
4. key word: difference
5. key word: reduced

Page 21

1. key word: change
$$\begin{array}{r} \$50 \\ - \$36 \\ \hline \mathbf{\$14} \end{array}$$

2. key word: difference
$$\begin{array}{r} \$12,635 \\ - \$\ 7,849 \\ \hline \mathbf{\$4,786} \end{array}$$

3. key words: more than
$$\begin{array}{r} 161 \text{ pounds} \\ - 104 \text{ pounds} \\ \hline \mathbf{57 \text{ pounds}} \end{array}$$

4. key word: decrease
$$\begin{array}{r} \$215 \\ - \$\ 46 \\ \hline \mathbf{\$169} \end{array}$$

5. key word: less
$$\begin{array}{r} 21 \text{ pounds} \\ - 17 \text{ pounds} \\ \hline \mathbf{4 \text{ pounds}} \end{array}$$

Pages 23–24

1. key words: and, altogether
operation: add
$$\begin{array}{r} 86 \text{ books} \\ + 53 \text{ books} \\ \hline \mathbf{139 \text{ books}} \end{array}$$

2. key words: more than
operation: subtract
$$\begin{array}{r} 31 \text{ rings} \\ - 15 \text{ rings} \\ \hline \mathbf{16 \text{ rings}} \end{array}$$

3. key word: total
operation: add
$$\begin{array}{r} 564 \text{ votes} \\ + 365 \text{ votes} \\ \hline \mathbf{929 \text{ votes}} \end{array}$$

4. key word: decrease
operation: subtract
$$\begin{array}{r} 3,421 \text{ people} \\ - 2,530 \text{ people} \\ \hline \mathbf{891 \text{ people}} \end{array}$$

5. key word: farther
operation: subtract
$$\begin{array}{r} 10,000 \text{ miles} \\ - 3,000 \text{ miles} \\ \hline \mathbf{7,000 \text{ miles}} \end{array}$$

6. key word: increased
operation: add
$$\begin{array}{r} \$830 \\ + 35 \\ \hline \mathbf{\$865} \end{array}$$

7. key words: rose, more than
operation: subtract
$$\begin{array}{r} \$542 \\ - 426 \\ \hline \mathbf{\$116} \end{array}$$

Pages 26–27

1. dropped
2. rose
3. increase
4. decrease
5. less
6. less
7. lowered
8. raised
9. more

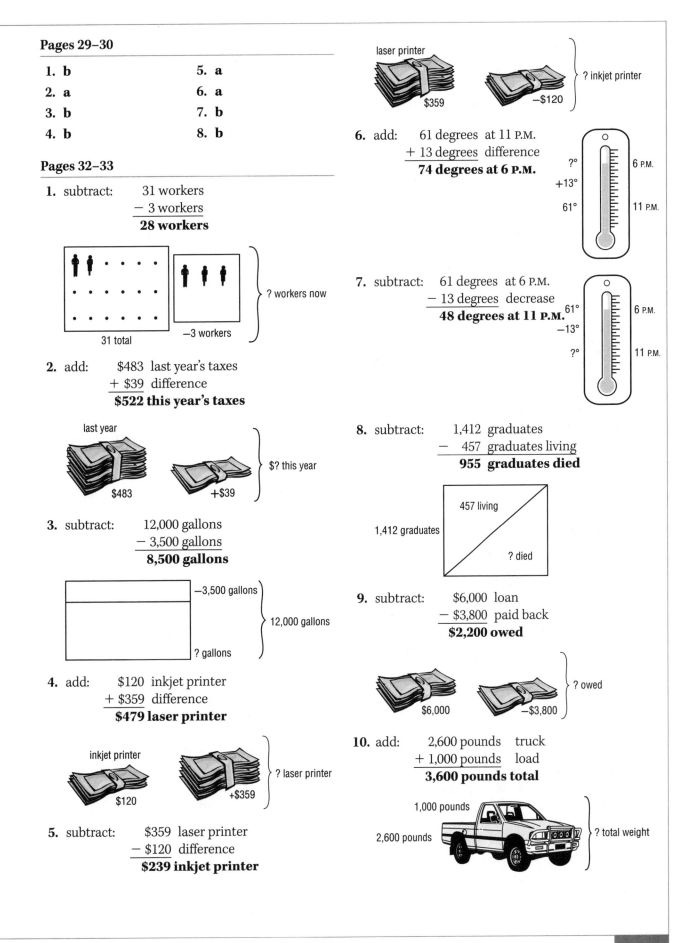

Pages 29–30

1. b	5. a
2. a	6. a
3. b	7. b
4. b	8. b

Pages 32–33

1. subtract:
31 workers
− 3 workers
28 workers

? workers now
31 total
−3 workers

2. add:
$483 last year's taxes
+ $39 difference
$522 this year's taxes

last year
$? this year
$483
+$39

3. subtract:
12,000 gallons
− 3,500 gallons
8,500 gallons

−3,500 gallons
12,000 gallons
? gallons

4. add:
$120 inkjet printer
+ $359 difference
$479 laser printer

inkjet printer
? laser printer
$120
+$359

5. subtract:
$359 laser printer
− $120 difference
$239 inkjet printer

laser printer
? inkjet printer
$359
−$120

6. add:
61 degrees at 11 P.M.
+ 13 degrees difference
74 degrees at 6 P.M.

?° 6 P.M.
+13°
61° 11 P.M.

7. subtract:
61 degrees at 6 P.M.
− 13 degrees decrease
48 degrees at 11 P.M.

61° 6 P.M.
−13°
?° 11 P.M.

8. subtract:
1,412 graduates
− 457 graduates living
955 graduates died

457 living
1,412 graduates
? died

9. subtract:
$6,000 loan
− $3,800 paid back
$2,200 owed

? owed
$6,000
−$3,800

10. add:
2,600 pounds truck
+ 1,000 pounds load
3,600 pounds total

1,000 pounds
2,600 pounds
? total weight

11. add and compare: 24 pounds
      + 42 pounds
      66 pounds

66 pounds is less than 70 pounds. The total weight did not exceed the weight limit.

12. add: 790,000 legal copies
       + 600,000 illegal copies
       1,390,000 copies

Pages 35–36

1. *necessary information:* 44-cent stamp, half dollar
word sentence: amount paid − price of stamp = amount of change
number sentence: 50 cents − 44 cents = amount of change

    50 cents
  − 44 cents
    6 cents

6 cents = amount of change

2. *necessary information:* 32 miles, 51 miles total
word sentence: miles driven − commuting distance = additional driving
number sentence: 51 miles − 32 miles = additional driving

    51 miles
  − 32 miles
    19 miles

19 miles = additional driving

3. *necessary information:* 134 tickets, 172 Saturday tickets
word sentence: ticket sales needed to break even − tickets sold = tickets to be sold
number sentence: 172 tickets − 134 tickets = tickets to be sold

    172 tickets
  − 134 tickets
    38 tickets

38 tickets = tickets to be sold

4. *necessary information:* $6,300, $1,460
word sentence: price of car − savings = loan
number sentence: $6,300 − $1,460 = loan

    $6,300
  − 1,460
    $4,840

$4,840 = loan

5. *necessary information:* 150 names, 119 names
word sentence: names needed − names collected = more names needed
number sentence: 150 names − 119 names = more names

    150 names
  − 119 names
    31 names

31 names = more names needed

6. *necessary information:* 47 pounds, 119 pounds
word sentence: new weight + weight loss = original weight
number sentence: 119 pounds + 47 pounds = original weight

    119 pounds
  + 47 pounds
    166 pounds

166 pounds = original weight

7. *necessary information:* $13, $168
word sentence: sale price + reduction = original price
number sentence: $168 + $13 = original price

    $168
  + 13
    $181

$181 = original price

8. *necessary information:* $865, $679
word sentence: original price − sale price = savings
number sentence: $865 − $679 = savings

    $865
  − 679
    $186

$186 = savings

9. *necessary information:* $2,130, $1,850
word sentence: dollars withheld − amount owed = refund
number sentence: $2,130 − $1,850 = refund

    $2,130
  − 1,850
    $280

$280 = refund

10. *necessary information:* 3,500 cars, 8,200 cars
    word sentence: new production + production cut = original production
    number sentence: 8,200 cars + 3,500 cars = original production

    $$\begin{array}{r} 8{,}200 \text{ cars} \\ + \ 3{,}500 \text{ cars} \\ \hline \mathbf{11{,}700 \ cars} \end{array}$$

    11,700 cars = original production

11. *necessary information:* $208,682, $327,991
    word sentence: sales − inventory cost = profit
    number sentence: $327,991 − $208,682 = profit

    $$\begin{array}{r} \$327{,}991 \\ - \ 208{,}682 \\ \hline \mathbf{\$119{,}309} \end{array}$$

    $119,309 = profit

12. *necessary information:* 1,423 gallons, 1,289 gallons
    word sentence: Bertha + Calico = total gallons
    number sentence: 1,423 gallons + 1,289 gallons = total gallons

    $$\begin{array}{r} 1{,}423 \text{ gallons} \\ + \ 1{,}289 \text{ gallons} \\ \hline \mathbf{2{,}712 \ gallons} \end{array}$$

    2,712 gallons = total gallons

13. *necessary information:* 72,070 seats, 58,682 people
    word sentence: total seats − attendance = empty seats
    number sentence:
    72,070 seats − 58,682 people = empty seats

    $$\begin{array}{r} 72{,}070 \text{ seats} \\ - \ 58{,}682 \text{ people} \\ \hline \mathbf{13{,}388 \ seats} \end{array}$$

    13,388 seats = empty seats

14. *necessary information:* 49 days, 56 days
    word sentence: spinach days + green beans days = total days
    number sentence: 49 days + 56 days = total days

    $$\begin{array}{r} 49 \text{ days} \\ + \ 56 \text{ days} \\ \hline \mathbf{105 \ days} \end{array}$$

    105 days = total days

Page 39

1. c	4. d
2. b	5. f
3. a	6. e

Page 41

1. f	4. d
2. e	5. a
3. c	6. b

Pages 42–43

Estimations may vary.

1. **b.**
   7.1 miles − 6.3 miles = **0.8 mile**
   estimation: 7 miles − 6 miles = **1 mile**

2. **a.**
   9.4 gallons + 14.7 gallons = **24.1 gallons**
   estimation: 9 gallons + 15 gallons = **24 gallons**

3. **a.**
   1.42 pounds + 0.98 pound = **2.40 pounds**
   estimation: 1 pound + 1 pound = **2 pounds**

4. **b.**
   200.15 mph − 198.7 mph = **1.45 mph**
   estimation: 200 mph − 199 mph = **1 mph**

5. **b.**
   9.1% − 7.9% = **1.2%**
   estimation: 9% − 8% = **1%**

Pages 44–45

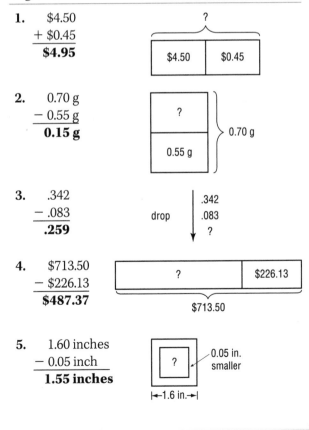

1. $$\begin{array}{r} \$4.50 \\ + \ \$0.45 \\ \hline \mathbf{\$4.95} \end{array}$$

2. $$\begin{array}{r} 0.70 \text{ g} \\ - \ 0.55 \text{ g} \\ \hline \mathbf{0.15 \ g} \end{array}$$

3. $$\begin{array}{r} .342 \\ - \ .083 \\ \hline \mathbf{.259} \end{array}$$

4. $$\begin{array}{r} \$713.50 \\ - \ \$226.13 \\ \hline \mathbf{\$487.37} \end{array}$$

5. $$\begin{array}{r} 1.60 \text{ inches} \\ - \ 0.05 \text{ inch} \\ \hline \mathbf{1.55 \ inches} \end{array}$$

6. $46.65
   + 63.35
   $110.00

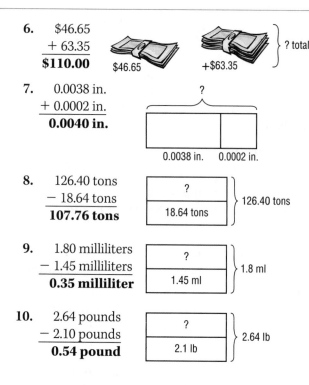

$46.65   +$63.35 } ? total

7. 0.0038 in.
   + 0.0002 in.
   0.0040 in.

?

0.0038 in.   0.0002 in.

8. 126.40 tons
   − 18.64 tons
   107.76 tons

?
18.64 tons

} 126.40 tons

9. 1.80 milliliters
   − 1.45 milliliters
   0.35 milliliter

?
1.45 ml

} 1.8 ml

10. 2.64 pounds
   − 2.10 pounds
   0.54 pound

?
2.1 lb

} 2.64 lb

Pages 46–47

1. *necessary information:* 0.6 ounce, 2.4 ounces
word sentence: new weight + reduction = original weight
number sentence: 2.4 ounces + 0.6 ounce = original weight

   2.4 ounces
   + 0.6 ounce
   3.0 ounces

3.0 ounces = original weight

2. *necessary information:* $7.38, $10.00
word sentence: amount paid − lunch cost = change
number sentence: $10.00 − $7.38 = change

   $10.00
   − $7.38
   $2.62

$2.62 = change

3. *necessary information:* 3.94 pounds, 4.68 pounds
word sentence: first chicken + second chicken = total weight
number sentence: 3.94 pounds + 4.68 pounds = total weight

   3.94 pounds
   + 4.68 pounds
   8.62 pounds

8.62 pounds = total weight

4. *necessary information:* 23,172.3 miles, 23,391.4 miles
word sentence: reading at end − reading at start = length of trip
number sentence:
23,391.4 miles − 23,172.3 miles = length of trip

   23,391.4 miles
   − 23,172.3 miles
   219.1 miles

219.1 miles = length of trip

5. *necessary information:* $1.40, $1.65
word sentence: bus + subway = 1-way trip
number sentence: $1.40 + $1.65 = 1-way trip

   $1.40
   + $1.65
   $3.05

$3.05 = 1-way trip

6. *necessary information:* $634.73, $595.99
word sentence: Massachusetts cost − New Hampshire cost = savings
number sentence: $634.73 − $595.99 = savings

   $634.73
   − $595.99
   $38.74

$38.74 = savings

7. *necessary information:* 14.36 seconds, 13.9 seconds
word sentence: first 100 meters + second 100 meters = total time
number sentence: 14.36 seconds + 13.9 seconds = total time

   14.36 seconds
   + 13.90 seconds
   28.26 seconds

28.26 seconds = total time

8. *necessary information:* 966 bottles, 50 bottles
word sentence: bottles by 1st line − number of bottles fewer = bottles by 2nd line
number sentence: 966 bottles − 50 bottles = bottles by 2nd line

   966 bottles
   − 50 bottles
   916 bottles

916 bottles = bottles by 2nd line

1. b.

$$28\tfrac{1}{2} = \quad 28\tfrac{2}{4} \text{ inches}$$
$$+\,31\tfrac{1}{4} = +\,31\tfrac{1}{4} \text{ inches}$$
$$\mathbf{59\tfrac{3}{4} \text{ inches}}$$

2. a.

$$1\tfrac{2}{3} \text{ cups}$$
$$+\,1\tfrac{1}{3} \text{ cups}$$
$$2\tfrac{3}{3} = \mathbf{3 \text{ cups}}$$

3. a.

$$71\tfrac{1}{4} = \quad 71\tfrac{1}{4} = \quad 70\tfrac{5}{4} \text{ pounds}$$
$$-\,62\tfrac{1}{2} = -\,62\tfrac{2}{4} = -\,62\tfrac{2}{4} \text{ pounds}$$
$$\mathbf{8\tfrac{3}{4} \text{ pounds}}$$

4. b.

$$23\tfrac{1}{4} = \quad 22\tfrac{5}{4} \text{ inches}$$
$$-\,18\tfrac{3}{4} = -\,18\tfrac{3}{4} \text{ inches}$$
$$4\tfrac{2}{4} \text{ inches} = \mathbf{4\tfrac{1}{2} \text{ inches}}$$

Page 50

Sample drawings. Actual diagrams may vary.

1.

$$4 = \quad 3\tfrac{8}{8} \text{ inches}$$
$$-\tfrac{5}{8} = -\tfrac{5}{8} \text{ inch}$$
$$\mathbf{3\tfrac{3}{8} \text{ inches}}$$

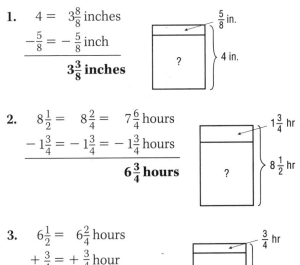

2.

$$8\tfrac{1}{2} = \quad 8\tfrac{2}{4} = \quad 7\tfrac{6}{4} \text{ hours}$$
$$-\,1\tfrac{3}{4} = -\,1\tfrac{3}{4} = -\,1\tfrac{3}{4} \text{ hours}$$
$$\mathbf{6\tfrac{3}{4} \text{ hours}}$$

3.

$$6\tfrac{1}{2} = \quad 6\tfrac{2}{4} \text{ hours}$$
$$+\tfrac{3}{4} = +\tfrac{3}{4} \text{ hour}$$
$$6\tfrac{5}{4} = \mathbf{7\tfrac{1}{4} \text{ hours}}$$

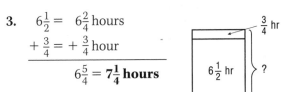

4.

$$2 = \quad 1\tfrac{2}{2} \text{ inches}$$
$$-\tfrac{1}{2} = -\tfrac{1}{2} \text{ inch}$$
$$\mathbf{1\tfrac{1}{2} \text{ inches}}$$

5.

$$2\tfrac{1}{2} = \quad 2\tfrac{3}{6} = \quad 1\tfrac{9}{6} \text{ cups}$$
$$-\,1\tfrac{2}{3} = -\,1\tfrac{4}{6} = -\,1\tfrac{4}{6} \text{ cups}$$
$$\mathbf{\tfrac{5}{6} \text{ cup}}$$

6.

$$\tfrac{3}{4} = \quad \tfrac{12}{16} \text{ inch}$$
$$-\tfrac{1}{16} = -\tfrac{1}{16} \text{ inch}$$
$$\mathbf{\tfrac{11}{16} \text{ inch}}$$

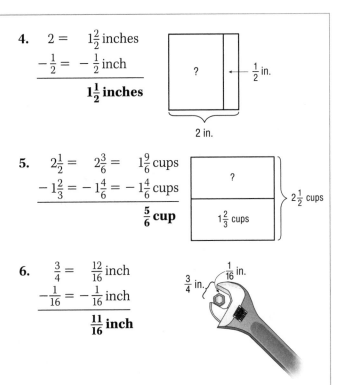

Page 51

1. bowl − rum = other ingredients

3 quarts − $1\tfrac{1}{4}$ quarts = other ingredients

$$3 = \quad 2\tfrac{4}{4} \text{ quarts}$$
$$-1\tfrac{1}{4} = -\,1\tfrac{1}{4} \text{ quarts}$$
$$\mathbf{1\tfrac{3}{4} \text{ quarts}} \text{ other ingredients}$$

2. orig. length − new length = amount taken off

$34\tfrac{1}{2}$ inches − $32\tfrac{3}{4}$ inches = amount taken off

$$34\tfrac{1}{2} = \quad 33\tfrac{6}{4} \text{ inches}$$
$$-32\tfrac{3}{4} = -\,32\tfrac{3}{4} \text{ inches}$$
$$\mathbf{1\tfrac{3}{4} \text{ inches}} \text{ amount taken off}$$

3. prior amount − amount sold = amount left

$$6\tfrac{1}{2} \text{ yards} - 3\tfrac{2}{3} \text{ yards} = \text{amount left}$$
$$6\tfrac{1}{2} = \quad 5\tfrac{9}{6} \text{ yards}$$
$$-3\tfrac{2}{3} = -\,3\tfrac{4}{6} \text{ yards}$$
$$\mathbf{2\tfrac{5}{6} \text{ yards}} \text{ amount left}$$

4. first week + second week = total wood

$$\tfrac{1}{8} \text{ cord} + \tfrac{1}{12} \text{ cord} = \text{total wood}$$

$$\tfrac{1}{8} = \tfrac{3}{24} \text{ cord}$$

$$+\tfrac{1}{12} = \tfrac{2}{24} \text{ cord}$$

$$\mathbf{\tfrac{5}{24} \text{ cord}} \text{ total wood}$$

5. total cloth − hem = length of drapes

$$62\tfrac{1}{2} \text{ inches} - \tfrac{3}{4} \text{ inch} = \text{length of drapes}$$

$$62\tfrac{1}{2} = 61\tfrac{6}{4} \text{ inches}$$

$$-\tfrac{3}{4} = -\tfrac{3}{4} \text{ inch}$$

$$\mathbf{61\tfrac{3}{4} \text{ inches}} \text{ length of drapes}$$

Page 53

Letters in equations and setup of equations may vary.

1. February − January = difference
4,348 hits − 2,917 hits = h
1,431 hits = h

2. amount in account before − cost of groceries = amount in account after
$n - \$78.62 = \$138.79$
$n = \$138.79 + \$78.62$
$n = \mathbf{\$217.41}$

3. calories of cheese pizza + calories of pepperoni = calories of pepperoni pizza

251 calories + c = 290 calories
− 251 calories = − 251 calories
c = **39 calories**

4. length of board − length needed = length left over
$3\tfrac{1}{2} \text{ feet} - 1\tfrac{3}{4} \text{ feet} = f$
$\mathbf{1\tfrac{3}{4} \text{ feet}} = f$

5. mg acetaminophen + mg aspirin + mg caffeine = mg active ingredients
250 mg + 250 mg + 65 mg = m
565 mg = m

6. current pile height − standard pile height = amount to be sheared off
$\tfrac{5}{8} \text{ in.} - \tfrac{7}{16} \text{ in.} = s$
$\mathbf{\tfrac{3}{16} \text{ in.}} = s$

Pages 54–56

1. b. *subtract:*
    55,572 nails
  − 1,263 nails
  54,309 nails

2. b. *subtract:* $\$75.62 - \$38.56 = \mathbf{\$37.06}$

3. d. *subtract:*
  29.1 miles per gallon
  − 26.2 miles per gallon
  2.9 miles per gallon

4. d. *add:* $\$1,800 + \$6,400 = \mathbf{\$8,200}$

5. c. *subtract:*
$\tfrac{1}{4} \text{ pound} = \tfrac{4}{16} \text{ pound}$
$-\tfrac{3}{16} \text{ pound} = -\tfrac{3}{16} \text{ pound}$
$\mathbf{\tfrac{1}{16} \text{ pound}}$

6. c. *subtract:*
  0.60 gram
  − 0.47 gram
  0.13 gram

7. a. *add:* $\$391,445 + \$528,555 = \mathbf{\$920,000}$

8. b. *add:*
$26\tfrac{3}{4} \text{ inches} = 26\tfrac{6}{8} \text{ inches}$
$+ 2\tfrac{3}{8} \text{ inches} = + 2\tfrac{3}{8} \text{ inches}$
$28\tfrac{9}{8} \text{ inches}$
$= \mathbf{29\tfrac{1}{8} \text{ inches}}$

9. e. *add:*
$\tfrac{5}{8} \text{ cord} = \tfrac{15}{24} \text{ cord}$
$+ \tfrac{1}{12} \text{ cord} = + \tfrac{2}{24} \text{ cord}$
$\mathbf{\tfrac{17}{24} \text{ cord}}$

10. e. *subtract:*
  5.15 tubes
  − 3.40 tubes
  1.75 tubes

11. b. *add:*
$\tfrac{1}{2} \text{ radioactivity} = \tfrac{8}{16} \text{ radioactivity}$
$+ \tfrac{7}{16} \text{ radioactivity} = + \tfrac{7}{16} \text{ radioactivity}$
$\mathbf{\tfrac{15}{16} \text{ radioactivity}}$

12. d. *add:*
  2.77 grams
  + 0.03 gram
  2.80 grams

Pages 57–58

1. times
2. per
3. total
4. twice
5. by
6. area
7. multiplied, times
8. per
9. per

Page 59

1. *necessary information:* $58, twice
   key word: twice

   $58
   $\times$ 2
   $116

2. *necessary information:* 4 sets, 6 strings per set
   key word: per

   6 strings per set
   $\times$ 4 sets
   24 strings

3. *necessary information:* 5 times a day, week
   key word: times

   5 times a day
   $\times$ 7 days (week)
   35 times

4. *necessary information:* $75 per hour, 3 hours
   key word: per

   $75 per hour
   $\times$ 3 hours
   $225

Page 60

1. shared, equally
2. per
3. average, each
4. shared, equally, each

Page 61

1. *necessary information:* 12 oz, 4 children
   key words: shared equally, each
   12 oz $\div$ 4 children

   3 oz
   4)$\overline{12}$

2. *necessary information:* 60-minute hockey game,
   3 equal periods
   key words: divided, equal (periods)
   60-minute hockey game $\div$ 3 equal periods

   20 minutes
   3)$\overline{60}$

3. *necessary information:* $156, 12 monthly
   payments
   key word: each
   $156 $\div$ 12 monthly payments

   $13
   12)$\overline{156}$

4. *necessary information:* 24 mints, 96 cents
   key word: each
   96 cents $\div$ 24 mints

   4 cents
   24)$\overline{96}$

5. *necessary information:* $3, $4,629
   key words: each, even
   $4,629 $\div$ $3

   1,543 tickets
   3)$\overline{4,629}$

Page 63

1. *division key words:* divided equally, each
   2 pieces
   4)$\overline{8}$

2. *division key words:* each, divided equally
   $18,000
   50)$\overline{900,000}$

3. *multiplication key words:* per, total
   35
   $\times$ 15
   $525

4. *multiplication key word:* quadruple
   47
   $\times$ 4
   $188

5. *division key words:* each, average
   45 points
   5)$\overline{225}$

6. *multiplication key word:* per
   38
   $\times$ 15
   570 miles

Page 66

1.
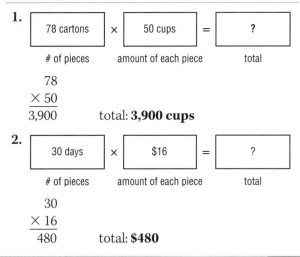

78 cartons	$\times$	50 cups	=	?
# of pieces		amount of each piece		total

   78
   $\times$ 50
   3,900   total: **3,900 cups**

2.

30 days	$\times$	$16	=	?
# of pieces		amount of each piece		total

   30
   $\times$ 16
   480   total: **$480**

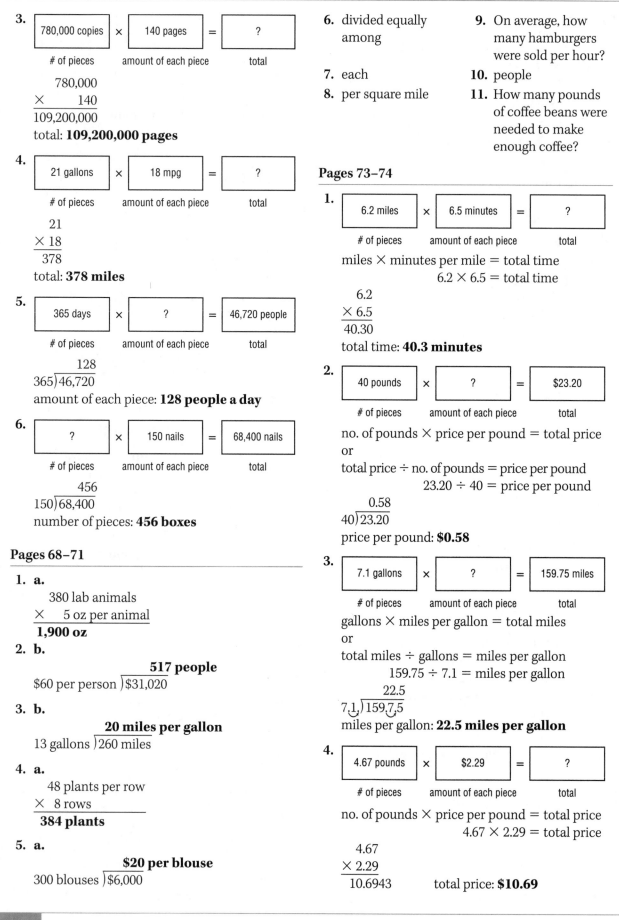

3.

780,000 copies	×	140 pages	=	?
# of pieces		amount of each piece		total

$$\begin{array}{r} 780{,}000 \\ \times\qquad 140 \\ \hline 109{,}200{,}000 \end{array}$$

total: **109,200,000 pages**

4.

21 gallons	×	18 mpg	=	?
# of pieces		amount of each piece		total

$$\begin{array}{r} 21 \\ \times\ 18 \\ \hline 378 \end{array}$$

total: **378 miles**

5.

365 days	×	?	=	46,720 people
# of pieces		amount of each piece		total

$$365\overline{)46{,}720}\quad 128$$

amount of each piece: **128 people a day**

6.

?	×	150 nails	=	68,400 nails
# of pieces		amount of each piece		total

$$150\overline{)68{,}400}\quad 456$$

number of pieces: **456 boxes**

Pages 68–71

1. a.
$$\begin{array}{r} 380 \text{ lab animals} \\ \times\quad 5 \text{ oz per animal} \\ \hline \mathbf{1{,}900 \text{ oz}} \end{array}$$

2. b.
$$\$60 \text{ per person }\overline{)\$31{,}020}\quad \mathbf{517 \text{ people}}$$

3. b.
$$13 \text{ gallons }\overline{)260 \text{ miles}}\quad \mathbf{20 \text{ miles per gallon}}$$

4. a.
$$\begin{array}{r} 48 \text{ plants per row} \\ \times\ 8 \text{ rows} \\ \hline \mathbf{384 \text{ plants}} \end{array}$$

5. a.
$$300 \text{ blouses }\overline{)\$6{,}000}\quad \mathbf{\$20 \text{ per blouse}}$$

6. divided equally among

7. each

8. per square mile

9. On average, how many hamburgers were sold per hour?

10. people

11. How many pounds of coffee beans were needed to make enough coffee?

Pages 73–74

1.

6.2 miles	×	6.5 minutes	=	?
# of pieces		amount of each piece		total

miles × minutes per mile = total time
6.2 × 6.5 = total time

$$\begin{array}{r} 6.2 \\ \times\ 6.5 \\ \hline 40.30 \end{array}$$

total time: **40.3 minutes**

2.

40 pounds	×	?	=	$23.20
# of pieces		amount of each piece		total

no. of pounds × price per pound = total price
or
total price ÷ no. of pounds = price per pound
23.20 ÷ 40 = price per pound

$$40\overline{)23.20}\quad 0.58$$

price per pound: **$0.58**

3.

7.1 gallons	×	?	=	159.75 miles
# of pieces		amount of each piece		total

gallons × miles per gallon = total miles
or
total miles ÷ gallons = miles per gallon
159.75 ÷ 7.1 = miles per gallon

$$7.1\overline{)159.7.5}\quad 22.5$$

miles per gallon: **22.5 miles per gallon**

4.

4.67 pounds	×	$2.29	=	?
# of pieces		amount of each piece		total

no. of pounds × price per pound = total price
4.67 × 2.29 = total price

$$\begin{array}{r} 4.67 \\ \times\ 2.29 \\ \hline 10.6943 \end{array}$$

total price: **$10.69**

5.

4 roommates	×	?	=	$372.36
# of pieces		amount of each piece		total

# of roommates × cost per roommate = total food bill

or

total food bill ÷ no. of roommates = cost per roommate

372.36 ÷ 4 = cost per roommate

$$\begin{array}{r} 93.09 \\ 4\overline{)372.36} \end{array}$$

cost per roommate: **$93.09 per roommate**

6.

8 hours	×	?	=	$118.56
# of pieces		amount of each piece		total

hours × hourly commission = total commission

or

total commission ÷ hours = hourly commission

118.56 ÷ 8 = hourly commission

$$\begin{array}{r} 14.82 \\ 8\overline{)118.56} \end{array}$$

hourly commission: **$14.82 per hour**

7.

?	×	$0.85	=	$167.45
# of pieces		amount of each piece		total

no. of passengers × cost per ride = $ collected

or

$ collected ÷ cost per ride = no. of passengers

167.45 ÷ 0.85 = no. of passengers

$$\begin{array}{r} 197. \\ 0.85\overline{)167.45} \end{array}$$

no. of passengers: **197 passengers**

8.

239 days	×	$3.65	=	?
# of pieces		amount of each piece		total

no. of days × cost per day = total cost

239 × 3.65 = total cost

$$\begin{array}{r} \$3.65 \\ \times\ \ 239 \\ \hline \$872.35 \end{array}$$ total cost: **$872.35**

Page 77

1. *necessary information:* $\frac{2}{3}$ of, 36 inches

fraction (of) × total precipitation = inches of rain

$\frac{2}{3} \times 36 = \frac{2}{\cancel{3}} \times \frac{\overset{12}{\cancel{36}}}{1} = $ **24 inches**

2. *necessary information:* $\frac{7}{8}$ of, 23,352 accidents

fraction (of) × total accidents = accidents in urban areas

$\frac{7}{8} \times 23,352 =$

$\frac{7}{\cancel{8}} \times \frac{\overset{2,919}{\cancel{23,352}}}{1} = $ **20,433 accidents**

3. *necessary information:* $7\frac{2}{3}$ pounds, 10 boxes

pounds per box × number of boxes = total pounds

$7\frac{2}{3} \times 10 = \frac{23}{3} \times \frac{10}{1} = \frac{230}{3} = $ **$76\frac{2}{3}$ pounds**

4. *necessary information:* $\frac{2}{3}$ can of, 12 days

cans per day × number of days = total cans

$\frac{2}{3} \times 12 = \frac{2}{\cancel{3}} \times \frac{\overset{4}{\cancel{12}}}{1} = $ **8 cans**

5. *necessary information:* $1\frac{1}{8}$ oz, $3\frac{1}{2}$ candy bars

oz per candy bar × no. of candy bars = total oz

$1\frac{1}{8} \times 3\frac{1}{2} = \frac{9}{8} \times \frac{7}{2} = \frac{63}{16} = $ **$3\frac{15}{16}$ oz**

6. *necessary information:* $\frac{1}{2}$ cup, $\frac{1}{2}$ of a load

fraction (of) × cups per load = detergent needed

$\frac{1}{2} \times \frac{1}{2}$ cup $= \frac{1}{2} \times \frac{1}{2} = $ **$\frac{1}{4}$ cup**

7. *necessary information:* $\frac{2}{5}$ of, $\frac{1}{4}$ pound

fraction (of) × weight of hamburger = amount of fat

$\frac{2}{5} \times \frac{1}{4}$ pound $= \frac{\cancel{2}}{5} \times \frac{\overset{1}{1}}{\cancel{4}} = $ **$\frac{1}{10}$ pound**

8. *necessary information:* 17,000 miles per hour, $2\frac{1}{2}$ hours

miles per hour × hours = miles traveled

$17,000 \times 2\frac{1}{2} = \frac{\overset{8,500}{\cancel{17,000}}}{1} \times \frac{5}{\cancel{2}} = $ **42,500 miles**

Page 79

1. *necessary information:* $22\frac{1}{2}$ in., $\frac{5}{8}$ in.

box depth ÷ thickness per book = no. of books

$22\frac{1}{2}$ in. deep ÷ $\frac{5}{8}$ in. per book $= \frac{\overset{9}{\cancel{45}}}{\cancel{2}} \times \frac{\overset{4}{\cancel{8}}}{\cancel{5}}$

$= $ **36 books**

2. *necessary information:* 13 people, $6\frac{1}{2}$ pounds

total amount of meat ÷ number of people = size of a portion

$6\frac{1}{2}$ lb meat ÷ 13 people $= \frac{\overset{1}{\cancel{13}}}{2} \times \frac{1}{\cancel{13}}$

$\qquad\qquad\qquad = \boldsymbol{\frac{1}{2}}$ **lb per person**

3. *necessary information:* $2\frac{1}{4}$ feet, $265\frac{1}{2}$ feet

total ribbon ÷ ribbon per book = number of books

$265\frac{1}{2}$ feet ÷ $2\frac{1}{4}$ feet $= \frac{531}{2} \div \frac{9}{4}$

$\qquad\qquad = \frac{\overset{59}{\cancel{531}}}{\underset{1}{\cancel{2}}} \times \frac{\overset{2}{\cancel{4}}}{\underset{1}{\cancel{9}}}$

$\qquad\qquad = \frac{59}{1} \times \frac{2}{1} = \boldsymbol{118}$ **books**

4. *necessary information:* $8\frac{1}{2}$ pounds, $\frac{1}{2}$ pound

total mashed potatoes ÷ potatoes per serving = number of servings

$8\frac{1}{2}$ pounds ÷ $\frac{1}{2}$ pound $= \frac{17}{\underset{1}{\cancel{2}}} \times \frac{\overset{1}{\cancel{2}}}{1} = \boldsymbol{17}$ **servings**

5. *necessary information:* $9\frac{3}{4}$ ounces, 3 equal servings

total ounces of peaches ÷ number of servings = size of each serving

$9\frac{3}{4}$ oz peaches ÷ 3 servings $= \frac{\overset{13}{\cancel{39}}}{4} \times \frac{1}{\underset{1}{\cancel{3}}} =$

$\frac{13}{4} = \boldsymbol{3\frac{1}{4}}$ **ounces per serving**

6. *necessary information:* 12 feet, $1\frac{1}{2}$-foot wide sections

total width ÷ width of section = number of sections

12 ft width ÷ $1\frac{1}{2}$ ft sections $= \frac{\overset{4}{\cancel{12}}}{1} \times \frac{2}{\underset{1}{\cancel{3}}}$

$\qquad\qquad\qquad = \boldsymbol{8}$ **sections**

Pages 80–81

1. *necessary information:* 12 hours, $\frac{3}{4}$ hour

total hours ÷ length of a session = number of sessions

12 hr total ÷ $\frac{3}{4}$ hr length session $= \frac{\overset{4}{\cancel{12}}}{1} \times \frac{4}{\underset{1}{\cancel{3}}}$

$\qquad\qquad\qquad\qquad = \boldsymbol{16}$ **sessions**

2. *necessary information:* $\frac{2}{3}$ hour, 30 cars

time per car × no. of cars = total hours

$\frac{2}{3}$ h × 30 no. of cars $= \frac{2}{\underset{1}{\cancel{3}}} \times \frac{\overset{10}{\cancel{30}}}{1} = \boldsymbol{20}$ **h**

3. *necessary information:* $\frac{2}{3}$ hour, 24 hours

total hours ÷ time per car = number of cars

24 hr total ÷ $\frac{2}{3}$ time per car $= \frac{\overset{12}{\cancel{24}}}{1} \times \frac{3}{\underset{1}{\cancel{2}}} = \boldsymbol{36}$ **cars**

4. *necessary information:* $\frac{2}{3}$ of, 26,148 ovens

fraction (of) × total ovens = defective ovens

$\frac{2}{3} \times 26{,}148$ ovens $= \frac{2}{\underset{1}{\cancel{3}}} \times \frac{\overset{8{,}716}{\cancel{26{,}148}}}{1}$

$\qquad\qquad = \boldsymbol{17{,}432}$ **ovens**

5. *necessary information:* 6 hikers, $4\frac{1}{2}$ pounds

total chocolate ÷ number of hikers = chocolate per hiker

$4\frac{1}{2}$ lb ÷ 6 hikers $= \frac{9}{2} \div \frac{6}{1} = \frac{\overset{3}{\cancel{9}}}{2} \times \frac{1}{\underset{2}{\cancel{6}}}$

$\qquad\qquad\qquad = \boldsymbol{\frac{3}{4}}$ **pound**

6. *necessary information:* $1\frac{1}{2}$ teaspoons, double

original number of teaspoons × 2 = doubled recipe

$1\frac{1}{2}$ tsp × 2 $= \frac{3}{2} \times \frac{2}{1} = \frac{6}{2} = \boldsymbol{3}$ **teaspoons**

7. On average, how much toner is used for each copy?

8. How much did she save on calls to her mother with her new phone card?

9. How much butter should she use for her new brownie recipe?

10. How much does a single fig bar weigh?

Pages 82–83

1. b. total weight ÷ weight of mayonnaise in each jar = number of jars

40 pounds ÷ $\frac{5}{8}$ pound per jar $= \frac{\overset{8}{\cancel{40}}}{1} \times \frac{8}{\underset{1}{\cancel{5}}}$

$\qquad\qquad\qquad = \boldsymbol{64}$ **jars**

2. c. total cost ÷ square feet = cost per sq ft

$\$20.80 ÷ 32$ square feet $= \boldsymbol{\$0.65}$

$\qquad \overset{\textstyle 0.65}{32\overline{)20.80}}$

3. b. fraction (of) × total raffle tickets = raffle sales needed

$\frac{1}{6} \times 3{,}000$ raffle tickets $=$

$\qquad \frac{1}{\underset{1}{\cancel{6}}} \times \frac{\overset{500}{\cancel{3{,}000}}}{1} = \boldsymbol{500}$ **raffle tickets**

4. b. fraction (of) × winner's share = trainer's share

$\frac{3}{5} \times \$17{,}490 = \frac{3}{\underset{1}{\cancel{5}}} \times \frac{\overset{3{,}498}{\cancel{17{,}490}}}{1} = \boldsymbol{\$10{,}494}$

5. d. total material ÷ material per apron = number of aprons

$7\frac{1}{3}$ yards ÷ $\frac{2}{3}$ yard/apron =

$\frac{22}{3} \div \frac{2}{3} = \frac{\overset{11}{\cancel{22}}}{\cancel{3}} \times \frac{\cancel{3}}{\cancel{2}} = $ **11 aprons**

6. d. total distance ÷ hours = miles per hour

263.1 miles ÷ 4.5 hours = **58.5 mi per hour**

$$58.46 = 58.5$$
$$4.5\,\overline{)263.1.}$$

7. c. total boxes ÷ boxes/carton = number of cartons

1,410 boxes ÷ 30 boxes/carton = **47 cartons**

$$\begin{array}{r} 47 \\ 30\,\overline{)1410} \end{array}$$

8. b. number of kilometers × mile per kilometer = number of miles

15 kilometers × 0.62 mile per kilometer = **9.3 miles**

$$\begin{array}{r} 15 \\ \times\ 0.62 \\ \hline 9.30 \end{array}$$

Page 84

2. $\dfrac{2 \text{ teachers}}{30 \text{ students}}$

3. $\dfrac{120 \text{ dollars}}{8 \text{ hours}}$

4. $\dfrac{38 \text{ miles}}{2 \text{ gallons}}$

5. $\dfrac{3 \text{ buses}}{114 \text{ commuters}}$

Page 87

1. $\dfrac{160 \text{ miles}}{5 \text{ hours}} = \dfrac{n \text{ miles}}{10 \text{ hours}}$

$5 \times n = 160 \times 10$

$5n = 1{,}600$

$n = \dfrac{1{,}600}{5}$

$n = $ **320 miles**

2. $\dfrac{12 \text{ cars}}{32 \text{ people}} = \dfrac{3 \text{ cars}}{n \text{ people}}$

$12 \times n = 32 \times 3$

$12n = 96$

$n = \dfrac{96}{12}$

$n = $ **8 people**

3. $\dfrac{n \text{ dollars}}{8 \text{ quarters}} = \dfrac{6 \text{ dollars}}{24 \text{ quarters}}$

$24 \times n = 8 \times 6$

$24n = 48$

$n = \dfrac{48}{24}$

$n = $ **2 dollars**

4. $\dfrac{42 \text{ pounds}}{n \text{ chickens}} = \dfrac{14 \text{ pounds}}{4 \text{ chickens}}$

$14 \times n = 42 \times 4$

$14n = 168$

$n = \dfrac{168}{14}$

$n = $ **12 chickens**

5. $\dfrac{28{,}928 \text{ people}}{8 \text{ doctors}} = \dfrac{n \text{ people}}{1 \text{ doctor}}$

$8 \times n = 1 \times 28{,}928$

$8n = 28{,}928$

$n = \dfrac{28{,}928}{8}$

$n = $ **3,616 people**

6. $\dfrac{\$24.39}{1 \text{ shirt}} = \dfrac{\$n}{6 \text{ shirts}}$

$1 \times n = \$24.39 \times 6$

$n = $ **\$146.34**

7. $\dfrac{\$47.85}{3 \text{ shirts}} = \dfrac{\$n}{10 \text{ shirts}}$

$3 \times n = 47.85 \times 10$

$3n = 478.50$

$n = \dfrac{478.50}{3}$

$n = $ **\$159.50**

8. $\dfrac{3 \text{ minutes}}{\frac{1}{2} \text{ mile}} = \dfrac{n \text{ minutes}}{5 \text{ miles}}$

$\frac{1}{2} \times n = 3 \times 5$

$\frac{1}{2}n = 15$

$n = 15 \div \frac{1}{2} = 15 \times 2$

$n = $ **30 minutes**

9. $\dfrac{575 \text{ passengers}}{n \text{ days}} = \dfrac{1{,}725 \text{ passengers}}{21 \text{ days}}$

$1{,}725 \times n = 575 \times 21$

$1{,}725n = 12{,}075$

$n = \dfrac{12{,}075}{1{,}725}$

$n = $ **7 days**

10. $\dfrac{7 \text{ blinks}}{\frac{1}{10} \text{ minute}} = \dfrac{n \text{ blinks}}{10 \text{ minutes}}$

$\frac{1}{10} \times n = 7 \times 10$

$\frac{1}{10}n = 70$

$n = 70 \times 10$

$n = $ **700 blinks**

Pages 89–90

1. *necessary information:* 7,800 people, 140,400 people; *labels for proportion:* $\frac{\text{shipments}}{\text{people}}$

$$\frac{1\text{ shipment}}{7,800\text{ people}} = \frac{n\text{ shipments}}{140,400\text{ people}}$$

$$7,800 \times n = 1 \times 140,400$$

$$7,800n = 140,400$$

$$n = \frac{140,400}{7,800}$$

$$n = \textbf{18 shipments}$$

2. *necessary information:* $340 an hour, 24 hours; *labels for proportion:* $\frac{\$}{\text{hours}}$

$$\frac{\$340}{1\text{ hour}} = \frac{\$n}{24\text{ hours}}$$

$$1 \times n = 340 \times 24$$

$$n = \textbf{\$8,160}$$

3. *necessary information:* 11 ounces, 28 cans; *labels for proportion:* $\frac{\text{ounces}}{\text{cans}}$

$$\frac{11\text{ ounces}}{1\text{ can}} = \frac{n\text{ ounces}}{28\text{ cans}}$$

$$1 \times n = 11 \times 28$$

$$n = \textbf{308 ounces}$$

4. *necessary information:* 52 words per minute, 26 minutes; *labels for proportion:* $\frac{\text{words}}{\text{minutes}}$

$$\frac{52\text{ words}}{1\text{ minute}} = \frac{n\text{ words}}{26\text{ minutes}}$$

$$1 \times n = 52 \times 26$$

$$n = \textbf{1,352 words}$$

5. *necessary information:* 3,960 Band-Aids, 180 school days; *labels for proportion:* $\frac{\text{Band-Aids}}{\text{school days}}$

$$\frac{3,960\text{ Band-Aids}}{180\text{ school days}} = \frac{n\text{ Band-Aids}}{1\text{ school day}}$$

$$180 \times n = 1 \times 3,960$$

$$180n = 3,960$$

$$n = \frac{3,960}{180}$$

$$n = \textbf{22 Band-Aids}$$

6. *necessary information:* 5,460 aspirins, 260 aspirins; *labels for proportion:* $\frac{\text{aspirins}}{\text{bottles}}$

$$\frac{260\text{ aspirins}}{1\text{ bottle}} = \frac{5,460\text{ aspirins}}{n\text{ bottles}}$$

$$260 \times n = 1 \times 5,460$$

$$260n = 5,460$$

$$n = \frac{5,460}{260}$$

$$n = \textbf{21 bottles}$$

7. *necessary information:* 126 tons, 3-ton loads; *labels for proportion:* $\frac{\text{tons}}{\text{loads}}$

$$\frac{3\text{ tons}}{1\text{ load}} = \frac{126\text{ tons}}{n\text{ loads}}$$

$$3 \times n = 1 \times 126$$

$$3n = 126$$

$$n = \frac{126}{3}$$

$$n = \textbf{42 loads}$$

8. *necessary information:* 18 feet, 3 feet in a yard; *labels for proportion:* $\frac{\text{feet}}{\text{yards}}$

$$\frac{3\text{ feet}}{1\text{ yard}} = \frac{18\text{ feet}}{n\text{ yards}}$$

$$3 \times n = 18 \times 1$$

$$3n = 18$$

$$n = \frac{18}{3}$$

$$n = \textbf{6 yards}$$

9. *necessary information:* 26 water chestnuts, 3 cans; *labels for proportion:* $\frac{\text{water chestnuts}}{\text{cans}}$

$$\frac{26\text{ water chestnuts}}{1\text{ can}} = \frac{n\text{ water chestnuts}}{3\text{ cans}}$$

$$1 \times n = 3 \times 26$$

$$n = \textbf{78 water chestnuts}$$

Page 92

1. *necessary information:* 25.4 millimeters per inch, 100-millimeter; *labels for proportion:* $\frac{\text{millimeters}}{\text{inches}}$

$$\frac{25.4\text{ millimeters}}{1\text{ inch}} = \frac{100\text{ millimeters}}{n\text{ inches}}$$

$$25.4 \times n = 1 \times 100$$

$$25.4n = 100$$

$$n = \frac{100}{25.4}$$

$$n = \textbf{3.94 inches}$$

2. *necessary information:* 2.2 pounds per kilogram, 36 kilograms; *labels for proportion:* $\frac{\text{pounds}}{\text{kilograms}}$

$$\frac{2.2\text{ pounds}}{1\text{ kilogram}} = \frac{n\text{ pounds}}{36\text{ kilograms}}$$

$$1 \times n = 2.2 \times 36$$

$$n = \textbf{79.2 pounds}$$

3. *necessary information:* 35.5 hours, $12.62 an hour; *labels for proportion:* $\frac{\text{hours}}{\$}$

$$\frac{35.5\text{ hours}}{\$n} = \frac{1\text{ hour}}{\$12.62}$$

$$1 \times n = 35.5 \times 12.62$$

$$n = \textbf{\$448.01}$$

4. *necessary information:* 1.09 yards per meter, 880 yards; *labels for proportion:* $\frac{\text{yards}}{\text{meters}}$

$$\frac{1.09 \text{ yards}}{1 \text{ meter}} = \frac{880 \text{ yards}}{n \text{ meters}}$$

$$1.09 \times n = 1 \times 880$$

$$1.09n = 880$$

$$n = \frac{880}{1.09}$$

$$n = \textbf{807.34 meters}$$

5. *necessary information:* \$38.00, \$4.20 per gallon; *labels for proportion:* $\frac{\$}{\text{gallons}}$

$$\frac{\$38}{n \text{ gallons}} = \frac{\$4.20}{1 \text{ gallon}}$$

$$4.20 \times n = 1 \times 38$$

$$4.2n = 38$$

$$n = \frac{38}{4.2}$$

$$n = \textbf{9.05 gallons}$$

6. *necessary information:* 1.61 kilometers, 55 miles per hour; *labels for proportion:*

$$\frac{\text{kilometers}}{\text{miles}} = \frac{\text{kilometers per hour}}{\text{miles per hour}}$$

(This proportion will work because the "per hour" appears on both top and bottom.)

$$\frac{1.61 \text{ kilometers}}{1 \text{ mile}} = \frac{n \text{ kilometers per hour}}{55 \text{ miles per hour}}$$

$$1 \times n = 1.61 \times 55$$

$$n = \textbf{88.55 kilometers per hour}$$

7. *necessary information:* 35.4 grams, 9,486 bars; *labels for proportion:* $\frac{\text{grams}}{\text{bars}}$

$$\frac{35.4 \text{ grams}}{1 \text{ Tiger Milk bar}} = \frac{n \text{ grams}}{9,486 \text{ Tiger Milk bars}}$$

$$1 \times n = 35.4 \times 9,486$$

$$n = \textbf{335,804.4 grams}$$

Page 93

1. *necessary information:* $\frac{1}{16}$ inch per slice, 2 inches; *labels for proportion:* $\frac{\text{slices}}{\text{inches}}$

$$\frac{1 \text{ slice}}{\frac{1}{16} \text{ inch}} = \frac{n \text{ slices}}{2 \text{ inches}}$$

$$\frac{1}{16} \times n = 1 \times 2$$

$$n = 2 \div \frac{1}{16}$$

$$n = \frac{2}{1} \times \frac{16}{1} = \textbf{32 slices}$$

2. *necessary information:* 1,460 loaves, $1\frac{3}{4}$ teaspoons per loaf; *labels for proportion:* $\frac{\text{teaspoons}}{\text{loaves}}$

$$\frac{1\frac{3}{4} \text{ teaspoons}}{1 \text{ loaf}} = \frac{n \text{ teaspoons}}{1,460 \text{ loaves}}$$

$$1 \times n = 1\frac{3}{4} \times 1,460$$

$$n = \frac{7}{4} \times \overset{365}{\cancel{1,460}} = \textbf{2,555 teaspoons}$$

3. *necessary information:* $\frac{7}{8}$ inch per book, 35 inches; *labels for proportion:* $\frac{\text{inches}}{\text{books}}$

$$\frac{\frac{7}{8} \text{ inch}}{1 \text{ book}} = \frac{35 \text{ inches}}{n \text{ books}}$$

$$\frac{7}{8} \times n = 1 \times 35$$

$$n = 35 \div \frac{7}{8}$$

$$n = \overset{5}{\cancel{35}} \times \frac{8}{\underset{1}{\cancel{7}}} = \textbf{40 books}$$

4. *necessary information:* $9\frac{2}{3}$ ounces per can, 16 cans; *labels for proportion:* $\frac{\text{cans}}{\text{ounces}}$

$$\frac{1 \text{ can}}{9\frac{2}{3} \text{ ounces}} = \frac{16 \text{ cans}}{n \text{ ounces}}$$

$$1 \times n = 9\frac{2}{3} \times 16$$

$$n = \frac{29}{3} \times \frac{16}{1} = \frac{464}{3}$$

$$n = \textbf{154}\frac{\textbf{2}}{\textbf{3}} \textbf{ ounces}$$

5. *necessary information:* 8 cups, $\frac{1}{4}$ cup per load; *labels for proportion:* $\frac{\text{cups}}{\text{loads}}$

$$\frac{8 \text{ cups}}{n \text{ loads}} = \frac{\frac{1}{4} \text{ cup}}{1 \text{ load}}$$

$$\frac{1}{4} \times n = 8 \times 1$$

$$n = 8 \div \frac{1}{4}$$

$$n = \frac{8}{1} \times \frac{4}{1} = \textbf{32 loads}$$

Page 95

1. *conversion:* 12 months = 1 year

$$\frac{12 \text{ months}}{1 \text{ year}} = \frac{30 \text{ months}}{n \text{ years}}$$

$$12 \times n = 1 \times 30$$

$$12n = 30$$

$$n = \frac{30}{12} = \frac{5}{2} = \textbf{2}\frac{\textbf{1}}{\textbf{2}} \textbf{ years}$$

2. *conversion:* 4 quarts = 1 gallon

$$\frac{4 \text{ quarts}}{1 \text{ gallon}} = \frac{n \text{ quarts}}{200 \text{ gallons}}$$

$$1 \times n = 4 \times 200$$

$$n = 800 \text{ quarts} = \textbf{800 bottles}$$

3. *conversion:* 1,000 meters = 1 kilometer

$$\frac{1{,}000 \text{ meters}}{1 \text{ kilometer}} = \frac{10{,}000 \text{ meters}}{n \text{ kilometers}}$$

$$1{,}000 \times n = 1 \times 10{,}000$$

$$1{,}000n = 10{,}000$$

$$n = \frac{10{,}000}{1{,}000} = \textbf{10 kilometers}$$

4. *conversion:* 2,000 pounds = 1 ton

$$\frac{2{,}000 \text{ pounds}}{1 \text{ ton}} = \frac{n \text{ pounds}}{\frac{1}{2} \text{ ton}}$$

$$1 \times n = \frac{\overset{1{,}000}{\cancel{2{,}000}}}{1} \times \frac{1}{\underset{1}{\cancel{2}}}$$

$$n = \textbf{1,000 pounds}$$

5. *conversion:* 32 ounces = 1 quart

$$\frac{32 \text{ ounces}}{1 \text{ quart}} = \frac{n \text{ ounces}}{3 \text{ quarts}}$$

$$1 \times n = 32 \times 3$$

$$n = \textbf{96 ounces}$$

6. *conversion:* 5,280 feet = 1 mile

$$\frac{5{,}280 \text{ feet}}{1 \text{ mile}} = \frac{29{,}028 \text{ feet}}{n \text{ miles}}$$

$$5{,}280 \times n = 1 \times 29{,}028$$

$$n = \frac{29{,}028}{5{,}280} = \textbf{5.5 miles}$$

Page 96

1.
$$\frac{\$12.60}{1 \text{ yard}} = \frac{\$n}{3\frac{1}{3} \text{ yards}}$$

$$1 \times n = \$12.60 \times 3\frac{1}{3}$$

$$n = \overset{4.20}{\cancel{12.60}} \times \frac{10}{\underset{1}{\cancel{3}}} = \textbf{\$42.00}$$

2.
$$\frac{13.5 \text{ pounds}}{4\frac{1}{2} \text{ years}} = \frac{n \text{ pounds}}{1 \text{ year}}$$

$$4\frac{1}{2} \times n = 13.5 \times 1$$

$$\frac{9}{2}n = 13.5$$

$$n = 13.5 \div \frac{9}{2}$$

$$n = 13.5 \times \frac{2}{9} = \frac{27.0}{9} = \textbf{3 pounds}$$

3.
$$\frac{2\frac{2}{3} \text{ pounds}}{\$3.25} = \frac{1 \text{ pound}}{\$n}$$

$$2\frac{2}{3} \times n = 3.25 \times 1$$

$$\frac{8}{3}n = 3.25$$

$$n = 3.25 \div \frac{8}{3}$$

$$n = 3.25 \times \frac{3}{8} = \frac{9.75}{8}$$

$$n = 1.218 \text{ or } \textbf{\$1.22}$$

4.
$$\frac{\$35.65}{1 \text{ cubic yard}} = \frac{\$n}{7\frac{1}{2} \text{ cubic yards}}$$

$$1 \times n = \$35.65 \times 7\frac{1}{2}$$

$$n = \textbf{\$267.38}$$

Page 97

1. *necessary information:* $\frac{1}{12}$ of an hour, 8 hours; *labels for proportion:* $\frac{\text{chickens}}{\text{hours}}$

$$\frac{1 \text{ chicken}}{\frac{1}{12} \text{ hour}} = \frac{n \text{ chickens}}{8 \text{ hours}}$$

$$\frac{1}{12} \times n = 1 \times 8$$

$$n = 8 \div \frac{1}{12}$$

$$n = 8 \times 12$$

$$n = \textbf{96 chickens}$$

2. *necessary information:* 8 blood samples, 60 minutes; *labels for proportion:* $\frac{\text{minutes}}{\text{blood samples}}$

$$\frac{60 \text{ minutes}}{8 \text{ blood samples}} = \frac{n \text{ minutes}}{1 \text{ blood sample}}$$

$$8 \times n = 1 \times 60$$

$$n = \frac{60}{8} = \frac{15}{2}$$

$$n = \textbf{7}\frac{\textbf{1}}{\textbf{2}} \textbf{ minutes}$$

3. *necessary information:* 1.6 kilometers, 26 miles; *labels for proportion:* $\frac{\text{kilometers}}{\text{miles}}$

$$\frac{1.6 \text{ kilometers}}{1 \text{ mile}} = \frac{n \text{ kilometers}}{26 \text{ miles}}$$

$$1 \times n = 1.6 \times 26$$

$$n = \textbf{41.6 kilometers}$$

4. *necessary information:* 6 feet an hour, 24 hours; *labels for proportion:* $\frac{\text{feet}}{\text{hours}}$

$$\frac{6 \text{ feet}}{1 \text{ hour}} = \frac{n \text{ feet}}{24 \text{ hours}}$$

$$1 \times n = 6 \times 24$$

$$n = \textbf{144 feet}$$

5. *necessary information:* 0.04 ounce per gram, 12-ounce can; *labels for proportion:* $\frac{\text{grams}}{\text{ounces}}$

$$\frac{1 \text{ gram}}{0.04 \text{ ounce}} = \frac{n \text{ grams}}{12 \text{ ounces}}$$

$$0.04 \times n = 1 \times 12$$

$$0.04n = 12$$

$$n = \frac{12}{0.04}$$

$$n = \textbf{300 grams}$$

6. *necessary information*: $3\frac{1}{4}$ pounds,
2 pumpkin pies, 10 pies;
labels for proportion: $\frac{\text{pounds}}{\text{pies}}$

$$\frac{3\frac{1}{4} \text{ pounds}}{2 \text{ pies}} = \frac{n \text{ pounds}}{10 \text{ pies}}$$

$$2 \times n = 10 \times 3\frac{1}{4}$$

$$2n = \frac{10}{1} \times \frac{13}{4}$$

$$2n = \frac{130}{4}$$

$$n = \frac{130}{4} \div 2$$

$$n = \frac{\overset{65}{\cancel{130}}}{4} \times \frac{1}{\underset{1}{\cancel{2}}}$$

$$n = \frac{65}{4}$$

$$n = \mathbf{16\frac{1}{4} \textbf{ pounds}}$$

7. *necessary information*: 68,000 gallons of water;
labels for proportion: $\frac{\text{gallons}}{\text{hours}}$

conversion: 1 day = 24 hours

$$\frac{68,000 \text{ gallons}}{1 \text{ hour}} = \frac{n \text{ gallons}}{24 \text{ hours}}$$

$$1 \times n = 68,000 \times 24$$

$$n = \mathbf{1,632,000 \textbf{ gallons}}$$

8. *necessary information*: \$0.40 per yard, $4\frac{1}{4}$ yards;
labels for proportion: $\frac{\$}{\text{yards}}$

$$\frac{\$0.40}{1 \text{ yard}} = \frac{\$n}{4\frac{1}{4} \text{ yards}}$$

$$1 \times n = \$0.40 \times 4\frac{1}{4}$$

$$n = \$\overset{.10}{\cancel{0.40}} \times \frac{17}{\underset{1}{\cancel{4}}}$$

$$n = \mathbf{\$1.70}$$

9. *necessary information*: $3\frac{1}{2}$ minutes;
labels for proportion: $\frac{\text{seconds}}{\text{minutes}}$

$$\frac{60 \text{ seconds}}{1 \text{ minute}} = \frac{n \text{ seconds}}{3\frac{1}{2} \text{ minutes}}$$

$$1 \times n = 60 \times 3\frac{1}{2}$$

$$n = \overset{30}{\cancel{60}} \times \frac{7}{\underset{1}{\cancel{2}}}$$

$$n = \mathbf{210 \textbf{ seconds}}$$

10. *necessary information*: 942 pages, 302,382 words;
labels for proportion: $\frac{\text{words}}{\text{pages}}$

$$\frac{302,382 \text{ words}}{942 \text{ pages}} = \frac{n \text{ words}}{1 \text{ page}}$$

$$942 \times n = 1 \times 302,382$$

$$942n = 302,382$$

$$n = \frac{302,382}{942}$$

$$n = \mathbf{321 \textbf{ words}}$$

11. *necessary information*: $\frac{1}{4}$ pound per burger,
50 pounds; *labels for proportion*: $\frac{\text{pounds}}{\text{hamburgers}}$

$$\frac{\frac{1}{4} \text{ pound}}{1 \text{ hamburger}} = \frac{50 \text{ pounds}}{n \text{ hamburgers}}$$

$$\frac{1}{4} \times n = 1 \times 50$$

$$n = 50 \times 4$$

$$n = \mathbf{200 \textbf{ hamburgers}}$$

12. *necessary information*: $\frac{2}{5}$ of all gallon-size
containers, 380 gallons;
labels for proportion: gallons

$$\frac{2}{5} = \frac{n \text{ gallons}}{380 \text{ gallons}}$$

$$5 \times n = 2 \times 380$$

$$5n = 760$$

$$n = \mathbf{152 \textbf{ gallons}}$$

Pages 98–99

1. multiplication
number of credits $\times$ cost of each credit =
total cost

2. division
total words $\div$ number of pages =
words per page

3. addition
workers now + workers laid off =
workers before

4. subtraction
sale price $-$ purchase price = profit

5. multiplication
molding per doorway $\times$ number of doorways =
total molding

6. subtraction
Thursday corn $-$ Wednesday corn = difference

7. division
total minutes $\div$ number of players =
minutes per player

8. division
1 gallon = 128 fluid ounces
total fl oz $\div$ no. of children = fl oz per child

9. addition
current amount = target amount + reduction

1. *necessary information labels:* ounce, pound
   answer label: ounces

   Are the labels different? yes
   new label: ounces

   ounces − ounces = ounces

   20 pounds = 320 ounces

   320 ounces
   − 8 ounces
   312 ounces

2. *necessary information labels:* feet, inches
   answer label: inches OR feet
   Are the labels different? yes
   new label: inches OR feet
   inches − inches = inches OR feet − feet = feet

   72 inches = 6 feet OR 2 feet = 24 inches

6 feet	OR	72 inches
− 2 feet		− 24 inches
4 feet		**48 inches**

3. *necessary information labels:* inches, inches
   answer label: inches

   Are the labels different? no

   inches + inches = inches

   $9\frac{1}{2}$ inches = $9\frac{2}{4}$ inches
   $+1\frac{3}{4}$ inches = $+1\frac{3}{4}$ inches
   $10\frac{5}{4}$ inches = $11\frac{1}{4}$ **inches**

4. $\frac{\text{sheets}}{\text{reams}} \times \text{reams} = \text{sheets}$

   $\frac{500 \text{ sheets}}{1 \text{ ream}} \times 10 \text{ reams} = \textbf{5,000 sheets}$

5. $\text{dollars} \div \frac{\text{dollars}}{\text{tickets}} = \text{tickets}$

   $\text{dollars} \times \frac{\text{tickets}}{\text{dollars}} = \text{tickets}$

   $\$5,220 \div \frac{\$9}{\text{ticket}} = ? \text{ tickets}$

   $\$5,220 \div 9 = \textbf{580 tickets}$

6. $\text{plants} \times \frac{\text{pounds}}{\text{plant}} = \text{pounds}$

   $14 \text{ plants} \times \frac{8 \text{ pounds}}{1 \text{ plant}} = \textbf{112 pounds}$

1. **b**   4. **c**

2. **d**   5. **d**

3. **a**

6. **e.** current year − age = birth year
   1998 − 86 = **1912**

7. **d.** meters per cable × number of cables =
   total meters
   35 meters × 7 cables = **245 meters**

8. **d.** old population + rise = new population
   894,943 people + 128,140 people =
   1,023,083 people

9. **a.** $\frac{\text{watts}}{\text{lights}} = \frac{\text{watts}}{\text{lights}}$
   $\frac{2}{1} = \frac{300}{n}$
   $2n = 300$
   $n = \frac{300}{2} = \textbf{150 lights}$

10. **c.** $\frac{\text{cables}}{\text{calls}} = \frac{\text{cables}}{\text{calls}}$
    $\frac{1}{12,500} = \frac{n}{87,500}$
    $12,500n = 87,500$
    $n = \frac{87,500}{12,500} = \textbf{7 cables}$

11. **e.** **not enough information given.** You do
    not know the tax rate.

12. **a.** total cloth ÷ number of looms =
    cloth per loom
    8,760 yards ÷ 60 looms = **146 yards**

13. **c.** month's sales − goal = extra sales
    37 SUVs − 20 SUVs = **17 SUVs**

Answers may vary.

14. What is the difference between the median
    household income of Maryland and Mississippi
    residents?

15. By how much was the price reduced?

16. How much water must he add to the muriactic
    acid to treat a concrete floor?

17. How many square feet of ventilation did Rosalia
    need to put in her attic?

18. How much sand must he add to the cement to
    make a batch of base coat stucco?

19. With this setting, what is the largest size the
    drilled hole might be?

20. What was the total cost of staying overnight at
    Motel 5?

1. c. total bill − tax = cost of sandwich alone
$$\$3.78 - \$0.18 = \mathbf{\$3.60}$$

2. c. United States − Russia = difference
309,187,390 people − 141,927,297 people = **167,260,093 people**

3. d. string ÷ length of pieces = number of pieces
12 feet ÷ $\frac{3}{4}$-foot piece = **16 pieces**

4. d. total amount for chicken ÷ price per pound = weight of chicken
$$\$8.70 \div \$1.95 = \mathbf{4.46\ pounds}$$

5. a. mother's material + daughter's material = total material
$$2\frac{1}{2}\ \text{yards} + \frac{7}{8}\ \text{yard} = \mathbf{3\frac{3}{8}\ yards}$$

6. c. peaches + plums = total weight
1.72 pounds + 0.9 pound = **2.62 pounds**

7. d. servings × grams per serving = total grams
11 × 0.26 gram = **2.86 grams**

8. b. cost of repair − amount deductible = insurance payment
$$\$1,125 - \$250 = \mathbf{\$875}$$

9. a. fraction (of) × paycheck = food cost
$$\frac{1}{3} \times \$414 = \mathbf{\$138}$$

10. d. total miles ÷ miles per check = total maintenance checks
96,000 miles ÷ 12,000 miles = **8 maintenance checks**

11. a. miles ÷ gallons = miles per gallon
283.1 miles ÷ 14.9 gallons = **19 miles per gallon**

12. b. normal depth + feet above normal = total depth
7 feet + 14 feet = **21 feet**

13. c. miles ÷ hours = average speed
$3,855 \div 3\frac{3}{4}$ hours = **1,028 miles per hour**

1. percent	**6.** percent
2. part	**7.** part
3. whole	**8.** percent
4. part	**9.** part
5. whole	**10.** whole

1. *whole:* $280
part: $84
percent: n (What was the percent?)
$$\frac{84}{280} = \frac{n}{100}$$
$$280 \times n = 84 \times 100$$
$$280n = 8,400$$
$$n = \frac{8,400}{280}$$
$$n = \mathbf{30\%}$$

2. *percent:* 58%
whole: 28,450 votes
part: n (How many votes did Marsha receive?)
$$\frac{n}{28,450} = \frac{58}{100}$$
$$100 \times n = 58 \times 28,450$$
$$100n = 1,650,100$$
$$n = \frac{1,650,000}{100}$$
$$n = \mathbf{16,501\ votes}$$

3. *percent:* 25%
part: $96,000
whole: n (How much aid was Metropolis receiving?)
$$\frac{96,000}{n} = \frac{25}{100}$$
$$25 \times n = 96,000 \times 100$$
$$25n = 9,600,000$$
$$n = \frac{9,600,000}{25}$$
$$n = \mathbf{\$384,000}$$

4. *percent:* 7%
part: $2,653
whole: n (What was his income?)
$$\frac{2,653}{n} = \frac{7}{100}$$
$$7 \times n = 2,653 \times 100$$
$$7n = 265,300$$
$$n = \frac{265,300}{7}$$
$$n = \mathbf{\$37,900}$$

5. *percent:* 60%
whole: 345,780
part: n (How many residents are African American?)
$$\frac{n}{345,780} = \frac{60}{100}$$
$$100 \times n = 345,780 \times 60$$
$$100n = 20,746,800$$
$$n = \frac{20,746,800}{100}$$
$$n = \mathbf{207,468\ African\ American\ residents}$$

6. *part:* $340
whole: $2,000
percent: n (What was the interest rate?)

$$\frac{340}{2,000} = \frac{n}{100}$$

$$2,000 \times n = 340 \times 100$$

$$2,000n = 34,000$$

$$n = \frac{34,000}{2,000}$$

$$n = \textbf{17\%}$$

7. *part:* n (What is the most calories from fat?)
whole: 2,400 calories
percent: 20%

$$\frac{n}{2,400} = \frac{20}{100}$$

$$100 \times n = 2,400 \times 20$$

$$100n = 48,000$$

$$n = \frac{48,000}{100}$$

$$n = \textbf{480 calories}$$

8. *part:* 3 grams fiber from serving of Cheerios
whole: n (recommended daily amount)
percent: 11%

$$\frac{3}{n} = \frac{11}{100}$$

$$11n = 300$$

$$n = \frac{300}{11}$$

$$n = \textbf{27 grams} \text{ (rounded to nearest gram)}$$

9. *part:* $12,500
whole: n (highest home purchase price that Noah and Juno can afford)
percent: 5%

$$5n = 1,250,000$$

$$n = \frac{1,250,000}{5}$$

$$n = \textbf{\$250,000}$$

10. c. 20% is the amount she saved. You are looking for the dollar amount she saved.

11. e. 720 dentists recommended Never Break. You are looking for the percent that recommended Never Break.

12. a. 60% of the legislators have to support a tax increase in order for it to pass. You are looking for the number of legislators needed to support a tax increase.

13. b. You are given that $75 is 30% off list price. You are looking for the list price.

14. d. 2% is the smallest percent of defective parts needed to shut down the production line. You are looking for the smallest number of defective parts that would result in the production line being shut down.

15. c. 59% is the percent of imported oil compared to total oil used in the United States. You are looking for the percent of daily consumption that depends on imported oil.

Pages 120–121

1. b. Since a state has thousands of workers, the percent is much more likely to be an estimate.

2. a. Sale prices will be exact to the nearest penny.

3. b. Predicting a future event like an earthquake is always an estimation.

4. b. It is likely that the actual background radiation level would be a fraction of a percent more or less than 35%.

5. a. 88% on a 50-question multiple-choice test is precisely 44 questions correct.

6. b. Cities have populations of thousands or more. It is very likely that the number of people living in poverty is not exactly 24%.

7. Since the report is probably rounded to the nearest percent, she might have received anywhere from 3,137 votes (at 64.5%) to 3,185 votes (1 less than 65.5%).

8. For the estimate to be accurate, the actual population must be between 134,210 (at 4.5%) and 135,494 (1 less than 5.5%).

9. The actual percent was slightly over 11.7%, which is close enough to 12% for the prediction to be considered accurate.

10. If the state says 3% of all tickets are winners, it is likely that the number will be exactly accurate, even with the large number of tickets. 600,000 winning tickets were printed.

11. The increase was about 23.1%. This would be considered a very accurate prediction.

Page 124

1. You are looking for the part. Therefore, you should multiply the total bill times the percent rate of the tip. $40.00 \times 0.15 = \textbf{\$6.00}$

2. You are looking for the part. Therefore, you should multiply the cost of the chain saw times the sales tax percent. $80 \times 0.05 =$ **$4.00**

3. You are looking for the percent. Therefore, you should divide the part (27 employees) by the whole (540 employees).

$$540\overline{)27.00}^{\,0.05}$$

Convert 0.05 to a percent: $0.05 \times 100 =$ **5%**

4. You are looking for the whole. Therefore, you should divide the part (1,500 customers who will visit the store) by the percent (2%).

$$.02\overline{)1,500.00}^{\,\textbf{75,000 circulars}}$$

5. You are looking for the percent. Therefore, you should divide the part (24 graduates) by the whole (30 graduates).

$$30\overline{)24.00}^{\,0.80}$$

Convert 0.80 to a percent: $0.80 \times 100 =$ **80%**

6. You are looking for the percent. Therefore, you should divide the part ($28,000), by the whole ($112,000).

$$112,000\overline{)28,000.00}^{\,0.25}$$

Convert 0.25 to a percent: $0.25 \times 100 =$ **25%**

Pages 126–127

1. *percent*: 4.5%
part: 90 homes
whole: n (How many homes were in the town?)
$$\frac{90}{n} = \frac{4.5}{100}$$
$$4.5 \times n = 90 \times 100$$
$$4.5n = 9{,}000$$
$$n = \frac{9{,}000}{4.5}$$
$$n = \textbf{2,000 homes}$$

2. *part*: $\frac{1}{10}$ pound of fat

whole: $\frac{3}{4}$ pound of boneless chicken breast

percent: n (What percent of the chicken breast was trimmed?)
$$\frac{\frac{1}{10}}{\frac{3}{4}} = \frac{n}{100}$$
$$\frac{3}{4} \times n = 100 \times \frac{1}{10}$$
$$\frac{3}{4}n = 10$$

$$n = 10 \div \frac{3}{4}$$
$$n = 10 \times \frac{4}{3} = \frac{40}{3} = 13\frac{1}{3}$$
$$n = \mathbf{13\tfrac{1}{3}\%}$$

$13\frac{1}{3}\%$ of the chicken breast was trimmed.

3. *percent*: 6.4%
whole: 800 visitors
part: n (How many visitors were visiting for the first time?)
$$\frac{n}{800} = \frac{6.4}{100}$$
$$100 \times n = 6.4 \times 800$$
$$100n = 5{,}120$$
$$n = \frac{5{,}120}{100}$$
$$n = \textbf{51.2}$$

51 visitors (51.2 rounded to the nearest person. You can't have 0.2 person visiting a Web site.)

4. *percent*: 8%
part: $49.76
whole: n (How much money?)
$$\frac{49.76}{n} = \frac{8}{100}$$
$$8 \times n = 49.76 \times 100$$
$$8n = 4{,}976$$
$$n = \frac{4{,}976}{8}$$
$$n = \textbf{\$622}$$

5. *whole*: $8.60
percent: 5%
part: n (How much was the tax?)
$$\frac{n}{8.60} = \frac{5}{100}$$
$$100 \times n = 8.60 \times 5$$
$$100n = 43$$
$$n = \frac{43}{100}$$
$$n = \textbf{\$0.43}$$

6. *whole*: $42.50
percent: 6%
part: n (Find the amount of tax.)
$$\frac{n}{42.50} = \frac{6}{100}$$
$$n \times 100 = 42.50 \times 6$$
$$100n = 255$$
$$n = \textbf{\$2.55}$$

7. *whole:* $15.95
percent: 7%
part: n (Find the amount of tax.)

$$\frac{n}{15.95} = \frac{7}{100}$$

$$n \times 100 = 15.95 \times 7$$

$$100n = 111.65$$

$$n = \mathbf{\$1.12}$$

8. *part:* $3.96
whole: $7.92
percent: n (By what percent?)

$$\frac{3.96}{7.92} = \frac{n}{100}$$

$$7.92 \times n = 3.96 \times 100$$

$$7.92n = 396$$

$$n = \mathbf{50\%}$$

9. *whole:* $12.80
percent: $12\frac{1}{2}\%$
part: n (How much did Juan save?)

$$\frac{n}{12.80} = \frac{12\frac{1}{2}}{100}$$

$$100 \times n = 12.80 \times 12\frac{1}{2}$$

$$100n = \overset{6.40}{\cancel{12.80}} \times \frac{25}{\underset{1}{\cancel{2}}}$$

$$100n = 160$$

$$n = \frac{160}{100}$$

$$n = \mathbf{\$1.60}$$

10. *percent:* 3.5%
part: 12,000 calls resulting in new subscribers
whole: n (How many calls?)

$$\frac{12,000}{n} = \frac{3.5}{100}$$

$$3.5n = 1,200,000$$

$$n = \mathbf{342{,}857} \text{ (round to nearest call}$$
$$\text{since you can't make a fraction of}$$
$$\text{a call)}$$

Pages 128–129

1. *part:* 3 out of
whole: 4 dentists
percent: n (What percent of all dentists?)

$$\frac{3}{4} = \frac{n}{100}$$

$$4 \times n = 3 \times 100$$

$$4n = 300$$

$$n = \frac{300}{4} = \mathbf{75\%}$$

2. *percent:* 40%
part: 112,492 people
whole: n (How many registered voters?)

$$\frac{112{,}492}{n} = \frac{40}{100}$$

$$40 \times n = 112{,}492 \times 100$$

$$40n = 11{,}249{,}200$$

$$n = \frac{11{,}249{,}200}{40} = \mathbf{281{,}230\ voters}$$

3. *part:* 11,500 fewer visitors
whole: 34,500 visitors
percent: n (What was the percent drop?)

$$\frac{11{,}500}{34{,}500} = \frac{n}{100}$$

$$34{,}500 \times n = 11{,}500 \times 100$$

$$34{,}500n = 1{,}150{,}000$$

$$n = \frac{1{,}150{,}000}{34{,}500} = \mathbf{33\tfrac{1}{3}\%}$$

4. *whole:* $49,600,000
percent: 8.6%
part: n (How much money did the company make?)

$$\frac{n}{49{,}600{,}000} = \frac{8.6}{100}$$

$$100 \times n = 8.6 \times 49{,}600{,}000$$

$$100n = 426{,}560{,}000$$

$$n = \frac{426{,}560{,}000}{100} = \mathbf{\$4{,}265{,}600}$$

5. *part:* 0.4 ounce
percent: $16\frac{2}{3}\%$
whole: n (What was the weight of its chocolate bar before the change?)

$$\frac{0.4}{n} = \frac{16\frac{2}{3}}{100}$$

$$16\tfrac{2}{3} \times n = 100 \times 0.4$$

$$16\tfrac{2}{3}n = 40$$

$$\frac{50}{3}n = 40$$

$$n = 40 \div \frac{50}{3}$$

$$n = \overset{4}{\cancel{40}} \times \frac{3}{\underset{5}{\cancel{50}}} = 4 \times \frac{3}{5}$$

$$n = \frac{12}{5} = \mathbf{2\tfrac{2}{5}\ or\ 2.4\ ounces}$$

6. *part:* 40
percent: 0.8%
whole: n (How many people over age 65?)

$$\frac{40}{n} = \frac{0.8}{100}$$

$$0.8 \times n = 40 \times 100$$

$$0.8n = 4{,}000$$

$$n = \frac{4{,}000}{0.8} = \mathbf{5{,}000\ people\ over\ age\ 65}$$

7. *percent:* 13%

whole: $41,694

part: n (How much did he pay?)

$$\frac{n}{41,694} = \frac{13}{100}$$

$$100 \times n = 13 \times 41,694$$

$$100n = 542,022$$

$$n = \mathbf{\$5,420.22}$$

8. *whole:* $17,548

percent: 7%

part: n (How much of a raise will he get?)

$$\frac{n}{17,548} = \frac{7}{100}$$

$$100 \times n = 17,548 \times 7$$

$$100n = 122,836$$

$$n = \frac{122,836}{100} = \mathbf{\$1,228.36}$$

9. *part:* $18

percent: 9%

whole: n (What had her week's salary been?)

$$\frac{18}{n} = \frac{9}{100}$$

$$9 \times n = 18 \times 100$$

$$9n = 1,800$$

$$n = \frac{1,800}{9} = \mathbf{\$200}$$

10. *part:* 506 scores

whole: 1,012 attempts

percent: n (What was his scoring percentage?)

$$\frac{506}{1,012} = \frac{n}{100}$$

$$1,012 \times n = 506 \times 100$$

$$1,012n = 50,600$$

$$n = \frac{50,600}{1,012} = \mathbf{50\%}$$

11. *part:* $\frac{3}{4}$ cup

percent: 30%

whole: n (How many cups of cereal must you eat?)

$$\frac{\frac{3}{4} \text{ cup}}{n} = \frac{30}{100}$$

$$\frac{3}{4} \times 100 = 30n$$

$$75 = 30n$$

$$\mathbf{2\frac{1}{2} \text{ cups}} = n$$

12. *whole:* $3\frac{3}{4}$ ounces

percent: n (What percent of vitamin D is provided?)

part: 1 ounce

$$\frac{1 \text{ ounce}}{3\frac{3}{4} \text{ ounce}} = \frac{n}{100}$$

$$1 \times 100 = 3\frac{3}{4} \times n$$

$$100 = \frac{15}{4}n$$

$$\mathbf{26.7\% \text{ or } 26\frac{2}{3}\%} = n$$

Pages 132–134

Information given in the problem is written under the appropriate place in the word sentence or proportion. Your wording may differ slightly.

1. *solution sentence:*

earnings − money taken out = take-home pay

  ($880)

missing information:

taxes + union dues = money taken out

($249)   ($26)

2. *solution sentence:*

starting money − money spent = money left

($71)

missing information:

lunch + gas = money spent

($5)   ($32)

3. *solution sentence:*

total cost ÷ number of dolls = cost per doll

($7,200)

missing information:

boxes × dolls per box = number of dolls

(30)   (8)

4. *solution sentence:*

starting balance + transactions = new balance

($394)

missing information:

deposit − check = transactions

($201)   ($187)

5. *solution sentence:*

cost of shirts + cost of skirt = total spent

            ($24)

missing information:

cost per shirt × no. of shirts = cost of shirts

($19)   (5)

6. *solution sentence:*

total amt ÷ monthly payments = amt of

            (24)   each payment

missing information:

loan + interest = total amt to pay

($4,600) ($728)

7. *proportion:*

$$\underset{(60)}{\overset{(10,000)}{\frac{\text{numbers entered}}{\text{minutes}}}} = \underset{(15)}{\frac{\text{numbers entered}}{\text{minutes}}}$$

8. *proportion:*

$$\underset{(2)}{\overset{(14)}{\frac{\text{grams per serving}}{\text{saturated fat per serving}}}} = \overset{(224)}{\frac{\text{grams per jar}}{\text{saturated fat per jar}}}$$

9. *solution sentence:*

$\$$ separate $-$ $\$$ combination $=$ savings
$\qquad$ ($\$5.29$)

missing information:

cheeseburger $+$ fries $+$ cola $=$ $\$$ separate
($\$2.89$) $\quad$ ($\$1.79$) $\;$ ($\$1.49$)

10. *solution sentence:*

ward delegates $+$ at-large $=$ total delegates
$\qquad$ ward delegates $+$ 5 $=$ total delegates

missing information:

wards $\times$ delegates $=$ ward delegates
$\qquad$ 8 $\times$ 4 $=$ 32 ward delegates

solution:

32 $+$ 5 $=$ **37 delegates**

11. *solution sentence:*

customer price $-$ cost $=$ profit
$\qquad$ $\$563$ $-$ cost $=$ profit

missing information:

parts $+$ labor $=$ cost
$\;$ $\$181$ $+$ $\$245$ $=$ $\$426$ cost

solution:

563 $-$ 426 $=$ **$\$137$ profit**

12. *solution sentence:*

raise $-$ new expenses $=$ net monthly increase
$\;$ $\$78$ $-$ new expenses $=$ net monthly increase

missing information:

daily increase $\times$ number of days $=$ new expenses

$\$2 \times 22 = \$44$ new expenses

solution:

$\$78 - \$44 =$ **$\$34$ net monthly increase**

13. $\dfrac{\text{people}}{\text{pounds}} = \dfrac{\text{people}}{\text{pounds}}$

$$\frac{6}{3} = \frac{8}{n}$$

$$6 \times n = 3 \times 8$$

$$6n = 24$$

$$n = \frac{24}{6} = \textbf{4 pounds}$$

14. $\dfrac{\text{last month }\$}{\text{last month kilowatts}} = \dfrac{\text{this month }\$}{\text{this month kilowatts}}$

$$\frac{90}{720} = \frac{n}{648}$$

$$720 \times n = 90 \times 648$$

$$720n = 58{,}320$$

$$n = \textbf{\$81}$$

last months bill $-$ this month's bill $=$ how much less

$\$90 - \$81 =$ **$\$9$ less**

15. *solution sentence:*

cu ft per day $\times$ # of days $=$ total cu ft
cu ft per day $\times$ # of days $=$ 1,728 cu ft

missing information:

cu ft per load $\times$ loads per day $=$ cu ft per day
$\quad$ 48 cu ft $\times$ 6 loads per day $=$
$\qquad\qquad\qquad\qquad\qquad$ 288 cu ft per day

solution:

288 cu ft per day $\times$ n days $=$ 1,728 cu ft

$$n = \frac{1{,}728}{288} = \textbf{6 workdays}$$

Pages 136–137

1. *solution sentence:*

total cost $\div$ total cookies $=$ cost per cookie
$\quad$ $\$24.40$ $\div$ total cookies $=$ cost per cookie

missing information:

cookies $\times$ number of boxes $=$ total cookies
$\qquad\qquad$ 20 $\times$ 6 $=$ 120 cookies

solution:

$\$24.40 \div 120 =$ **$\$0.20$ per cookie,** rounded to the nearest penny

2. *proportion:*

$$\frac{\text{reduction}}{\text{total price}} = \frac{\text{percent}}{100}$$

$$\frac{n}{400} = \frac{30}{100}$$

$$100n = 12{,}000$$

$$n = 120$$

original price $-$ reduction $=$ sale price
$\qquad$ $\$400 - \$120 =$ **$\$280$**

3. *solution sentence:*

pears $+$ fruit cocktail $=$ total fruit

$\qquad$ pears $+ 17\frac{1}{2} =$ total fruit

missing information:

cans $\times$ contents $=$ ounces of pears

$\qquad$ $5 \times 9\frac{3}{4} = 48\frac{3}{4}$ ounces

solution:

$48\frac{3}{4} + 17\frac{1}{2} = \mathbf{66\frac{1}{4}}$ **ounces**

4. *solution sentence:*

$\;$ total savings $-$ cost $=$ net savings
total savings $-$ $\$2.49$ $=$ net savings

missing information:
savings × months = total savings

$$\$3.40 \times 12 = \$40.80$$

solution:

$$\$40.80 - \$2.49 = \mathbf{\$38.31\ net\ savings}$$

5. *solution sentence:*
original weight − loss = cooked weight

$$\tfrac{1}{4}\text{ pound} - \text{loss} = \text{cooked weight}$$

missing information:
weight × fraction = loss

$$\tfrac{1}{4} \times \tfrac{1}{3} = \tfrac{1}{12}\text{ loss}$$

solution:

$$\tfrac{1}{4} - \tfrac{1}{12} = \mathbf{\tfrac{1}{6}\ pound}$$

6. *solution sentence:*
dinner + tax = total cost

$$\$24 + \text{tax} = \text{total cost}$$

missing information:
dinner × percent = tax

$$\$24 \times 0.06 = \$1.44$$

solution:

$$\$24 + \$1.44 = \mathbf{\$25.44\ total}$$

7. *proportion:*

$$\frac{\text{discount}}{\text{original price}} = \frac{\text{percent}}{100}$$

$$\frac{n}{39.50} = \frac{30}{100}$$

$$100n = 1185$$

$$n = \$11.85$$

original price − discount = sale price

$$\$39.50 - \$11.85 = \mathbf{\$27.65}$$

8. *solution sentence:*
payment ÷ months = monthly payment

$$\text{payment} \div 12 = \text{monthly payment}$$

missing information:
amount − down payment = payment

$$\$610.60 - \$130 = \$480.60$$

solution:

$$\$480.60 \div 12 = \mathbf{\$40.05\ monthly\ payment}$$

9. *solution sentence:*
price per bottle − cost per bottle = profit per bottle

$$\$2.10 - \text{cost per bottle} = \text{profit per bottle}$$

missing information:
cost ÷ bottles = cost per bottle

$$\$57 \div 30 = \$1.90$$

solution:

$$\$2.10 - \$1.90 = \mathbf{\$0.20\ profit}$$

10. *proportion:*

$$\frac{\text{part}}{\text{whole}} = \frac{\text{percent}}{100}$$

$$\frac{\text{profit}}{57} = \frac{n\text{ percent}}{100}$$

missing information:
bottles × profit per bottle = total profit

$$30 \times \$0.20 = \$6$$

solution:

$$\frac{6}{57} = \frac{n}{100}$$

$$57 \times n = 6 \times 100$$

$$57n = 600$$

$$n = \frac{600}{57} = \mathbf{10.5\%}$$

11. $\dfrac{\text{ounces salt}}{\text{ounces stew}} = \dfrac{\text{ounces salt}}{\text{ounces stew}}$

$$\frac{0.2}{256} = \frac{n}{12}$$

$$256 \times n = 0.2 \times 12$$

$$256n = 2.4$$

$$n = \mathbf{0.0093\ ounce\ salt}$$

$$(\text{or } n = \mathbf{0.01\ ounce\ salt})$$

12. *solution sentence:*
total material ÷ material/outfit = no. of outfits

$$70\text{ yards} \div \text{material/outfit} = \text{no. of outfits}$$

missing information:
top + shirt = material/outfit

$$\tfrac{2}{3} + 1\tfrac{1}{4} = 1\tfrac{11}{12}\text{ yards}$$

solution:

$$70\text{ yards} \div 1\tfrac{11}{12}\text{ yard} = \mathbf{36\ outfits}$$

(discard remainder since you are asked for complete outfits)

Page 139

1. $20 - 9 = 11$

2. $20 - 6 + 3 = 14 + 3 = 17$

3. $157.8 - 5.7 - 1.1 = 152.1 - 1.1 = 151$

4. $19.2 + 7.8 = 27$

5. $10 \div 5 = 2$

6. $36 \div 3 + 6 = 12 + 6 = 18$

7. $36 \div 3 + 6 = 12 + 6 = 18$

8. $90 - 10 - 8 = 80 - 8 = 72$

9. $1.8 \div 0.02 - 2 = 90 - 2 = 88$

10. $56 - 24.5 = 31.5$

Answer Key   209

Page 141

1. **d**

2. **a**

3. **b**

4. **c**

5. **e**

Pages 143–144

1. **b.**
   $46 + 48 = 94$
   $94 - 81 = \textbf{\$13}$

2. **e.**
   $5 \times 256 = 1,280$
   $1,280 \div 8 = \textbf{160 fluid ounces}$

3. **b.**
   $85 \times 80 = 6,800$
   $6,800 \div 100 = \textbf{\$68}$

4. **b.**
   $6.79 \div 32 = 0.21$ (rounded to the nearest cent)
   $11.98 \div 64 = 0.19$ (rounded to the nearest cent)
   $0.21 - 0.19 = \textbf{\$0.02 per fluid ounce}$

Page 146–147

1. **c.**

2. **e.**

3. **d.**

4. **a.**

5. **b.**

6. **d.**

7. **b.**

8. **a.**

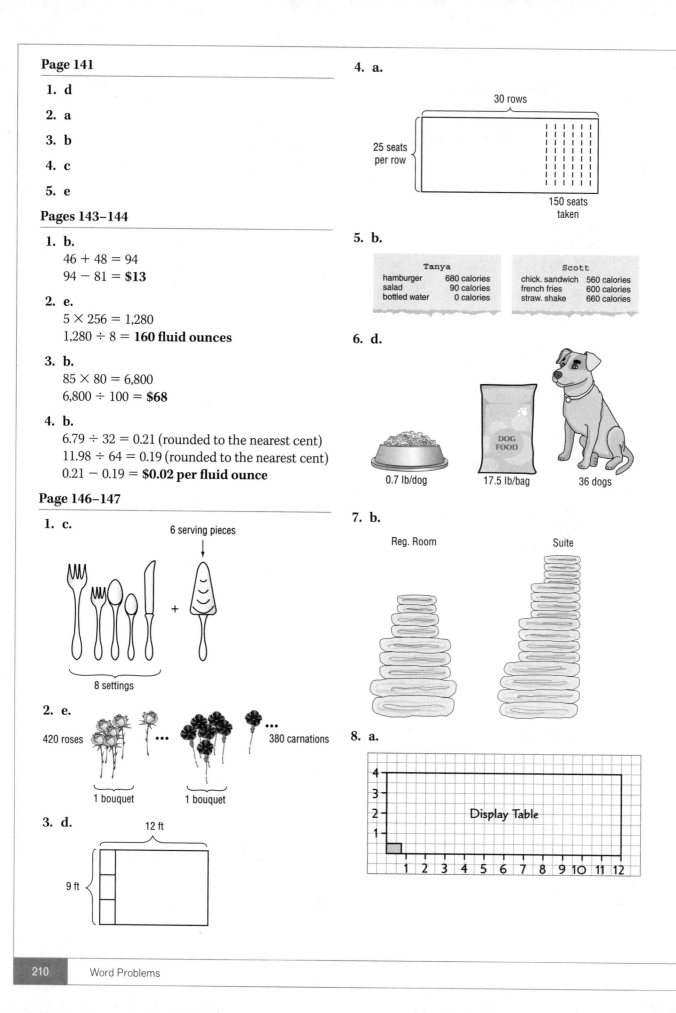

Page 149

1. floor size ÷ tile size = number of tiles
   floor size ÷ 81 = number of tiles
   conversion:
   (144 square inches in a square foot)
   $$\frac{\text{square inches}}{1 \text{ square foot}} = \frac{n \text{ square inches}}{\text{square feet}}$$
   $$\frac{144}{1} = \frac{n}{54}$$
   $$n = 144 \times 54 = 7{,}776 \text{ square inches}$$
   solution:
   7,776 ÷ 81 = **96 tiles**

2. welds per hour × hours = total welds
   20 × 9 = **180 total welds**

3. highway ÷ reflector distance = number of reflectors
   highway ÷ 528 = reflectors
   conversion: (1 mile = 5,280 feet)
   $$\frac{\text{feet}}{\text{mile}} = \frac{\text{feet}}{\text{mile}}$$
   $$\frac{5{,}280}{1} = \frac{n}{46}$$
   $$n = 242{,}880 \text{ feet}$$
   solution:
   242,880 ÷ 528 = **460 reflectors**
   (Another acceptable answer is 461 reflectors. This would depend on whether the first or the second reflector was at the first 528-foot mark.)

4. gallons × pints per gallon = total people
   gallons × 8 = total people
   (There are 8 pints in a gallon.)
   missing information:
   gallons − donations = gallons needed
   48 − 26 = 22 gallons needed
   solution:
   22 × 8 = **176 people**

5. total coal ÷ pounds per power plant = power plants
   total coal ÷ 4,000 = number of power plants
   conversion: (There are 2,000 pounds in a ton.)
   $$\frac{\text{pounds}}{1 \text{ ton}} = \frac{\text{pounds}}{\text{ton}}$$
   $$\frac{2{,}000}{1} = \frac{n}{38}$$
   $$n = 76{,}000 \text{ pounds}$$
   solution:
   76,000 ÷ 4,000 = **19 power plants**

6. $$\frac{\text{numbers}}{\text{time}} = \frac{\text{numbers}}{\text{time}}$$
   (There are 60 minutes in an hour.)
   $$\frac{463}{5} = \frac{n}{60}$$

$$5n = 27{,}780$$
$$n = \frac{27{,}780}{5} = \textbf{5{,}556 numbers}$$

7. ice cream ÷ serving = number of people
   ice cream ÷ 4 = number of people
   conversion: (1 quart contains 32 ounces.)
   $$\frac{\text{quarts}}{\text{ounces}} = \frac{\text{quarts}}{\text{ounces}}$$
   $$\frac{1}{32} = \frac{12}{n}$$
   $$n = 384 \text{ ounces}$$
   solution:
   384 ÷ 4 = **96 people**

Page 151

1. *necessary information:* 103,912 students, 4,657 students, 1,288 students original
   students − change = new total
   103,912 − change = new total
   missing information:
   left − enrolled = change
   4,657 − 1,288 = 3,369 students
   solution:
   103,912 − 3,369 = **100,543 students**

2. *necessary information:* 103,912 students, 1,288 students
   original + new enrollees = total students
   103,912 + 1,288 = **105,200 students**

3. *necessary information:* 54 degrees, 27 degrees, 19 degrees
   high temp. − drop = midnight temp.
   high temp. − 19 = midnight temp.
   missing information:
   first + rise = high temp.
   54 + 27 = 81 degree high temp.
   solution:
   81 − 19 = **62 degrees midnight temp.**

4. *necessary information:* $548, 52 weeks
   weekly income × no. of weeks = yearly income
   548 × 52 = **$28,496**

5. *necessary information:* 3 paperbacks, 2 magazines, $9.95, $3.50, $50
   money paid − cost of purchases = change
   $50 − cost of purchases = change
   missing information: price of books + price of magazines = cost of purchases
   9.95 × 3 + 3.50 × 2 = $36.85
   solution:
   50.00 − 36.85 = **$13.15**

6. *necessary information:* 20%, $4.80

cost − amount of discount = final price

$4.80 − amount of discount = final price

missing information:

cost × percent = amount of discount

$4.80 × 0.20 = \$0.96$

solution:

$4.80 − 0.96 = $ **\$3.84 final price**

7. necessary information: 28 grams, 3 grams

$$\frac{\text{grams fiber}}{\text{total grams}} = \frac{n\%}{100}$$

$$\frac{3 \text{ grams}}{28 \text{ grams}} = \frac{n}{100}$$

$$28 \times n = 3 \times 100$$

$$n = \frac{300}{28}$$

$n = 10.7\%$ (rounded to nearest tenth of a percent)

8. necessary information: 40 hours, $6.40 an hour, $186

salary + tips = total earned

salary + $186 = total earned

missing information:

hours × $ per hour = salary

40 hours × $6.40 per hour = $256

solution:

256 + 186 = **\$442**

Page 153

1. apples + cantaloupe = total

apples + $2.88 = total

missing information:

$$\frac{\text{cost}}{\text{apples}} = \frac{\text{cost}}{\text{apples}}$$

$$\frac{3.76}{12} = \frac{n}{7}$$

$$12n = 26.32$$

$$n = 26.32 \div 12 = \$2.19$$

solution:

$2.19 + $2.88 = **\$5.07**

2. bill − discount = new bill

$96.80 − discount = new bill

missing information:

bill × percent = discount

96.80 × 0.06 = $5.81

solution:

$96.80 − $5.81 = **\$90.99**

3. reduced calories − breakfast = rest of day

reduced calories − 797 = rest of day

missing information:

original − (% × calories) = reduced calories

4,200 − (0.28 × 4,200) = reduced calories

4,200 − 1,176 = 3,024 calories

solution:

3,024 − 797 = **2,227 calories**

4. salary + commission = pay

$150 + commission = pay

missing information:

percent × sales over 400 = commission

0.06 × (6,160 − 400) = commission

0.06 × 5,760 = $345.60

solution:

$150 + $345.60 = **\$495.60**

5. money spent ÷ gallons = price per gallon

338 ÷ gallons = price per gallon

missing information:

miles ÷ miles per gallon = gallons

3,627 ÷ 31 = 117 gallons

solution:

338 ÷ 117 = **\$2.89 per gallon**

Pages 154–155

1. c. daily miles × days = total miles

daily miles × 5 = total miles

missing information:

to work + back + delivery = daily miles

7 + 7 + 296 = 310 miles

solution:

310 × 5 = **1,550 miles**

2. a. travel miles + miles at work = daily miles

14 + 296 = **310 miles**

3. c. total ÷ students = individual cost

total ÷ 15 = individual cost

missing information:

books + materials = total

420 + 300 = $720

solution:

720 ÷ 15 = **\$48 individual cost**

4. e. small rooms' capacities + main room capacity = total capacity

small rooms' capacities + 94 = total capacity

missing information:

rooms × capacity = small rooms' capacities

4 × 28 = 112 people

solution:

112 + 94 = **206 people**

5. b. players × teams = players before change
$36 \times 8 = \textbf{288 players}$ before change

6. d. players on new rosters × teams = total after change
players on new rosters × 8 = total after change
missing information:
players − reduction = players on new rosters
$36 - 3 = 33$ players
solution:
$33 \times 8 = \textbf{264 players}$

7. c. cheese cost + apples cost = total cost
missing information:
pounds × cost = cheese cost
$2.36 \times 2.58 = \$6.09$
pounds × cost = apples cost
$4 \times 0.89 = \$3.56$
solution:
$\$6.09 + \$3.56 = \textbf{\$9.65}$

8. c. front-loading − top-loading = additional loads
missing information:
cups ÷ amount = front-loading
$6 \div \frac{1}{4} = 6 \times 4 = 24$ front loads
cups ÷ amount = top-loading
$6 \div \frac{1}{3} = 6 \times 3 = 18$ top loads
solution:
$24 - 18 = \textbf{6 loads}$

Pages 156–159, Posttest A

1. $\dfrac{\text{cocoa}}{\text{chocolate}} = \dfrac{\text{cocoa}}{\text{chocolate}}$

$\dfrac{3 \text{ tablespoons}}{1 \text{ ounce}} = \dfrac{n \text{ tablespoons}}{12 \text{ ounces}}$

$1 \times n = 3 \times 12$
$n = \textbf{36 tablespoons}$

2. $\dfrac{\text{part}}{\text{whole}} = \dfrac{\text{percent}}{100}$

$\dfrac{25 \text{ square inches}}{400 \text{ square inches}} = \dfrac{n\%}{100}$

$400 \times n = 25 \times 100$
$n = 2{,}500 \div 400 = \mathbf{6\frac{1}{4}\%}$

3. total children − children not in public school = children in public school
$5{,}372 \text{ children} - 1{,}547 \text{ children} = \textbf{3,825 children}$

4. $\dfrac{\text{pounds}}{\text{bushel}} = \dfrac{\text{pounds}}{\text{bushel}}$

$\dfrac{48 \text{ pounds}}{1 \text{ bushel}} = \dfrac{12 \text{ pounds}}{n \text{ bushels}}$

$48 \times n = 1 \times 12$
$n = 12 \div 48 = \mathbf{\frac{1}{4} \textbf{ bushel}}$

5. hot dog + soda = total spent
$\$2.50 + \$1.25 = \textbf{\$3.75}$

6. $\dfrac{\text{part}}{\text{whole}} = \dfrac{\text{percent}}{100}$

missing information: $80\% - 70\% = 10\%$

$\dfrac{n}{48{,}657} = \dfrac{10}{100}$

$100 \times n = 10 \times \$48{,}657$
$n = \textbf{\$4,865.70}$

7. $\dfrac{\text{liquid}}{\text{corn syrup}} = \dfrac{\text{liquid}}{\text{corn syrup}}$

$\dfrac{\frac{1}{4} \text{ cup liquid}}{1 \text{ cup corn syrup}} = \dfrac{n \text{ cup liquid}}{1\frac{1}{2} \text{ cups corn syrup}}$

$1 \times n = \frac{1}{4} \times 1\frac{1}{2}$
$n = \frac{1}{4} \times \frac{3}{2} = \mathbf{\frac{3}{8} \textbf{ cup}}$

8. charge per member × number of members = total collected
missing information: dues + magazine = charge per member
$20 + 15 = \$35$
$\$35 \times 13{,}819 \text{ members} = \textbf{\$483,665}$

9. $\dfrac{\text{part}}{\text{whole}} = \dfrac{\text{percent}}{100}$

$\dfrac{\$26{,}000}{n} = \dfrac{8}{100}$

$8 \times n = 26{,}000 \times 100$
$n = 2{,}600{,}000 \div 8 = \textbf{\$325,000}$

10. jeans + swim suit + sweater = total spent on clothes
$\$134 + \$98 + \$74 = \textbf{\$306}$

11. original value − value lost = 5-year value
missing information:
$\frac{1}{3}$ original value + $\frac{1}{4}$ original value = value lost
$\frac{1}{3} \times 12{,}900 + \frac{1}{4} \times 12{,}900 =$ value lost
$4{,}300 + 3{,}225 = \$7{,}525$
$12{,}900 - 7{,}525 = \textbf{\$5,375}$

12. feet in a mile × number of miles = total number of feet
missing information:
feet in a yard × yards in a mile = feet in a mile
$3 \text{ feet} \times 1{,}760 \text{ yards} = 5{,}280 \text{ feet}$
$5{,}280 \text{ feet} \times 5 \text{ miles} = \textbf{26,400 feet}$

13. original weight − weight lost = new weight
$172\frac{1}{2} \text{ pounds} - 47\frac{3}{4} \text{ pounds} = \mathbf{124\frac{3}{4} \textbf{ pounds}}$

14. *missing information:* You need to know how many miles are in a lap.

15. profit ÷ number of women = profit per woman
missing information:
earnings − expenses = profit
336,460 − 123,188 = $213,272
$213,272 ÷ 4 women = **$53,318**

16. $\dfrac{\text{part}}{\text{whole}} = \dfrac{\text{percent}}{100}$

$\dfrac{n}{768} = \dfrac{20}{100}$

$100 \times n = 20 \times 768$

$n = 15{,}360 \div 100 =$ **$153.60**

17. voters for incumbent × voters for challenger = decided voters
$\dfrac{1}{3} + \dfrac{1}{4} = \dfrac{7}{12}$ **of the voters**

18. $ per hour × number of hours = total earned
$14.20 × 17.5 hours = **$248.50**

19. *missing information:* You need to know the base price of the car.

20. total homes ÷ no. of people = homes per person
948 homes ÷ 12 people = **79 homes**

21. total bill − air-conditioning = bill without air conditioner
116.29 − 59.00 = **$57.29**

22. cost of socks + cost of towels = total spent
missing information:
cost per sock × no. of socks = cost of socks
$3.79 × 3 socks = $11.37
cost per towel × no. of towels = cost of towels
$4.69 × 4 towels = $18.76
$11.37 + $18.76 = **$30.13**

23. container size ÷ serving size = no. of servings
9 pounds ÷ $\dfrac{1}{16}$ = **144 sundaes**

24. dinner + movie + babysitter = cost of evening
$28.43 + $9.50 + $25.00 = **$62.93**

25. total weight ÷ weight per book = no. of books
34.2 pounds ÷ 0.6 pound = **57 books**

Pages 163–165

1. c. grams carbs OJ + grams carbs milk shake + grams carbs cola = total grams
22 grams + 60 grams + 34 grams =
116 grams carbohydrates

2. d. proportion:
$\dfrac{12 \text{ fluid ounces}}{10 \text{ grams fat}} = \dfrac{18 \text{ fluid ounces}}{n \text{ grams fat}}$
$12n = 180$
$n =$ **15 grams fat**

3. d. calories steak − calories salmon = difference in calories
914 calories − n calories = difference in calories
proportion:
$\dfrac{8 \text{ ounces}}{412 \text{ calories}} = \dfrac{12 \text{ ounces}}{n \text{ calories}}$
$8n = 4{,}944$
$n = 618$ calories
914 calories − 618 calories = **296 calories**

4. Answers may vary. Make reasonable estimations.
Steak:
$\dfrac{12 \text{ ounces}}{914 \text{ calories}} = \dfrac{n \text{ ounces}}{600 \text{ calories}}$
$900n = 7{,}200$ (914 rounded to 900)
$n =$ **8 ounces**

Hamburger on bun: $\dfrac{5 \text{ ounces}}{405 \text{ calories}} = \dfrac{n \text{ ounces}}{600 \text{ calories}}$
$400n = 3{,}000$ (405 rounded to 400)
$n =$ **7.5 ounces**
Fried chicken breast: $\dfrac{6 \text{ ounces}}{442 \text{ calories}} = \dfrac{n \text{ ounces}}{600 \text{ calories}}$
$450n = 3{,}600$ (442 rounded to 450)
$n =$ **8 ounces**
Spaghetti, tomato sauce:
$\dfrac{10 \text{ ounces}}{246 \text{ calories}} = \dfrac{n \text{ ounces}}{600 \text{ calories}}$
$250n = 6{,}000$ (246 rounded to 250)
$n =$ **24 ounces**

Broiled salmon: $\dfrac{8 \text{ ounces}}{412 \text{ calories}} = \dfrac{n \text{ ounces}}{600 \text{ calories}}$
$400n = 4{,}800$ (412 rounded to 400)
$n =$ **12 ounces**

5. fat from 10-ounce sirloin steak − fat from other entree = how much less fat
$\dfrac{12 \text{ ounces}}{57 \text{ grams fat}} = \dfrac{10 \text{ ounces}}{n \text{ grams fat}}$
$12n = 570$
$n =$ **47.5 grams fat** from 10-ounce sirloin steak
hamburger on bun: $\dfrac{5 \text{ ounces}}{21 \text{ grams fat}} = \dfrac{10 \text{ ounces}}{n \text{ grams fat}}$
$5n = 210$
$n = 42$ grams fat
47.5 − 42 = **5.5 grams less fat**
fried chicken breast: $\dfrac{6 \text{ ounces}}{22 \text{ grams fat}} = \dfrac{10 \text{ ounces}}{n \text{ grams fat}}$

$6n = 220$
$n = 36.7$ grams from fat (nearest tenth)
$47.5 - 36.7 = $ **10.8 grams less fat**
spaghetti, tomato sauce: $47.5 - 1 = $
46.5 grams less fat
salmon: $\dfrac{8 \text{ ounces}}{17 \text{ grams fat}} = \dfrac{10 \text{ ounces}}{n \text{ grams fat}}$
$8n = 170$
$n = 21.25$
$47.5 - 21.25 = $ **26.25 grams less fat**

6. Hamburger is ground up steak. According to the chart, steak has no carbohydrates. Therefore, the carbohydrates in a hamburger on bun come from the bun.

7. **e.** Broiled salmon gives the best combination of high protein and low fat. The steak has more protein but far more fat. The spaghetti with tomato sauce has much less fat but also much less protein. The hamburger on bun and the fried chicken have both less protein and more fat than the salmon.

8. **c.** and **e.**
jogging for 40 min + swimming for 30 min + bicycling for 2 hours ($10 \text{ cal/min} \times 40 \text{ min}$) + ($4 \text{ cal/min} \times 30 \text{ min}$) + ($4 \text{ cal/min} \times 120 \text{ min}$)
$400 \text{ cal} + 120 \text{ cal} + 480 \text{ cal} = $ **1,000 calories**
cross-country skiing for 2 hours + aerobic dancing for 35 min
($6 \text{ cal/min} \times 120 \text{ min}$) + ($8 \text{ cal/min} \times 35 \text{ min}$)
$720 \text{ cal} + 280 \text{ cal} = $ **1,000 calories**

9. **b.** and **d.**
running for 25 min + walking for 45 min + bicycling for 25 min
($19 \text{ cal/min} \times 25 \text{ min}$) + ($5 \text{ cal/min} \times 45 \text{ min}$) + ($4 \text{ cal/min} \times 25 \text{ min}$)
$475 \text{ cal} + 225 \text{ cal} + 100 \text{ cal} = $ **800 calories**

aerobic dancing for 40 min + tennis for 40 min + jogging for 24 min
($8 \text{ cal/min} \times 40 \text{ min}$) + ($6 \text{ cal/min} \times 40 \text{ min}$) + ($10 \text{ cal/min} \times 24 \text{ min}$)
$320 \text{ cal} + 240 \text{ cal} + 240 \text{ cal} = $ **800 calories**

10. **a.**
recreational volleyball for 90 min + tennis for 55 min
($3 \text{ cal/min} \times 90 \text{ min}$) + ($6 \text{ cal/min} \times 55 \text{ min}$)
$270 \text{ cal} + 330 \text{ cal} = $ **600 calories**

11. Answers will vary. Make sure that total minutes are at least 60.
walking for 100 minutes

$5 \text{ cal/min} \times 100 \text{ min} = 500 \text{ calories}$
swimming for 40 minutes + jogging for 34 minutes ($4 \text{ cal/min} \times 40 \text{ min}$) + ($10 \text{ cal/min} \times 34 \text{ min}$) $160 \text{ cal} + 340 \text{ cal} = 500 \text{ calories}$

12. Answers will vary. Make sure that total minutes is exactly 60.
aerobic dancing for 30 minutes + swimming for 30 minutes
($8 \text{ cal/min} \times 30 \text{ min}$) $\times$ ($4 \text{ cal/min} \times 30 \text{ min}$)
$240 \text{ cal} + 120 \text{ cal} = 360 \text{ calories}$

cross-country skiing for 40 minutes + walking for 20 minutes
($6 \text{ cal/min} \times 40 \text{ min}$) + ($5 \text{ cal/min} \times 20 \text{ min}$)
$240 \text{ cal} + 100 \text{ cal} = $ **340 calories**

Pages 167

1. length $\times$ width = area
$75 \text{ ft} \times 35 \text{ ft} = A$
2,625 sq ft $= A$

2. $\dfrac{\text{pounds}}{\text{square feet}} = \dfrac{\text{pounds}}{\text{square feet}}$
$\dfrac{40 \text{ pounds}}{1,000 \text{ square feet}} = \dfrac{n \text{ pounds}}{\text{area of lawn}}$
length $\times$ width = area
$80 \text{ ft} \times 30 \text{ ft} = 2,400 \text{ square feet}$
$\dfrac{40}{1,000} = \dfrac{n}{2,400}$
$1,000n = 96,000$
$n = $ **96 pounds**

3. labor rate $\times$ hours = total charge
\$35 per hour $\times$ hours = total charge
missing information: work hours + travel hours = total hours
$5 \text{ hours} + \frac{1}{2} \text{ hour} = 5\frac{1}{2} \text{ hours}$
\$35 per hour $\times 5\frac{1}{2} = $ total charge
$35 \times \frac{11}{2} = $ **\$192.50**

4. wholesale price of fertilizer bag + amount of markup = retail price of fertilizer bag
\$28 per bag + amount of markup = retail price of fertilizer bag
missing information: $\dfrac{\%\text{markup}}{100} = \dfrac{\$n}{\text{wholesale price}}$
$\dfrac{60}{100} = \dfrac{n}{28}$
$100 \times n = 60 \times 28$
$100n = 1,680$
$n = \$16.80$
\$28 + \$16.80 = retail price of bag
\$44.80 = retail price of bag

5. perimeter of yard $-$ length of house = feet of fencing

perimeter of yard − 100 = feet of fencing
missing information: perimeter of yard =
2 × length + 2 × width
$P = 2 \times 120 + 2 \times 50 = 240 + 100 = 340$ feet
340 feet − 100 feet = feet of fencing
    240 feet = feet of fencing

6. linear feet of fencing × cost per linear foot =
total cost of fencing
316 feet × $2.60 per foot = total cost of fencing
    $821.60 = total cost of fencing

7. difference in salary per hour × no. of hours =
difference in pay
difference in salary per hour × 7 = difference
in pay
missing information: Kendricks hourly pay −
Marquis' hourly pay = difference in pay
per hour
11.62 − 9.78 = difference in pay per hour
    $1.84 = difference in pay per hour
$1.84 × 7 = difference in pay
    $12.88 = difference in pay

8. hourly rate × number of hours = total earned
    $15.40 × 32.5 = total earned
    $500.50 = total earned

Pages 168–169

1. *missing information:* $\frac{\text{percent}}{100} = \frac{\text{markup}}{\text{wholesale price}}$
$\frac{50}{100} = \frac{m}{\$35}$
$100 \times m = 50 \times 35$
$100m = 1750$
$m = \$17.50$
$35 + 17.50 = $ retail price
    $52.50 = retail price

2. $\frac{\%}{100} = \frac{\text{savings}}{\text{marked price}}$
$\frac{20}{100} = \frac{s}{\$84}$
$100 \times s = 20 \times 84$
$100s = 1{,}680$
$s = \textbf{\$16.80}$

3. original price − amount off = clearance price
    $96 − amount off = clearance price
missing information: $\frac{\%}{100} = \frac{\text{reduction}}{\text{original price}}$
$\frac{65}{100} = \frac{r}{\$96}$
$100 \times r = 65 \times 96$
$100r = 6240$
$r = \$62.40$
$96 − 62.40 = $ clearance price
    $33.60 = clearance price

4. rent & utilities + salaries + inventory & misc =
total to break even
$892.80 + $5,193.85 + $3,645.00 = total
    $9,731.65 = total

5. payment − total price of items = change
missing information: jacket + gloves = total
price
$68.99 + $22.49 = total price
    $91.48 = total price
    $100 − $91.48 = change
    $8.52 = change

6.   scarf + belt + vest + shipping = total
$12.95 + $17.49 + $36.89 + shipping = total
    $67.33 + shipping = total
Use the chart to determine shipping: $67.33 is
between $38.00 and $74.99, so the shipping is
$12.95.
$67.33 + $12.95 = total price
    $80.28 = total price

7. pants suit + tops + crew sox + shipping =
total price
$119.65 + (3 × $27.85) + (4 × 3.49) + shipping
= total price
$119.65 + $83.55 + $13.96 + shipping =
total price
$217.16 + shipping = total price
Use the chart to determine shipping: $217.16
is between $200 and $299.99, so the shipping
is $39.95.
$217.16 + $39.95 = total price
    $257.11 = total price

8. $\frac{\text{payment to PayFriend}}{\text{total price}} = \frac{2\%}{100}$
$\frac{n}{\$80.28} = \frac{2}{100}$
$100n = 160.56$
$n = \textbf{\$1.61}$

9. $\frac{\text{payment to PayFriend}}{\text{total price}} = \frac{2\%}{100}$
$\frac{n}{\$257.11} = \frac{2}{100}$
$100n = 514.22$
$n = \textbf{\$5.14}$

Pages 170–171

1. tablets per dose × no. of doses =
total tablets
missing information:
$\frac{\text{total dose}}{\text{mg Tylenol per tablet}} = $ tablets per dose

$$\frac{650 \text{ mg}}{325 \text{ mg}} = 2 \text{ tablets per dose}$$

$$\frac{\text{hours per day}}{\text{hours between doses}} = \text{number of doses}$$

$$\frac{24 \text{ hours}}{4 \text{ hours}} = 6 \text{ doses}$$

$$2 \times 6 = \text{total tablets}$$

$$\mathbf{12} = \text{total tablets}$$

2. $\dfrac{\text{liters}}{\text{day}} = \dfrac{\text{milliliters}}{\text{hour}}$

$$\frac{2 \text{ liters}}{24 \text{ hours}} = \frac{m \text{ ml}}{1 \text{ hour}}$$

$$24 \times m = 2 \times 1$$

$$m = \frac{2}{24} = 0.083 \text{ liter per hour}$$

conversion: $\dfrac{1 \text{ liter}}{1,000 \text{ milliliters}} = \dfrac{0.083 \text{ liter per hour}}{m \text{ ml per hour}}$

$$\mathbf{m = 83 \text{ ml per hour}}$$

3. $\dfrac{\text{ordered dose}}{\text{dose per tablet}} = \dfrac{n \text{ tablets}}{1 \text{ tablet}}$

$$0.25 \times n = 0.125 \times 1$$

$$\frac{0.125}{0.25} = \frac{1}{2} \textbf{ tablet per day}$$

4. $\dfrac{\text{total strips}}{\text{strips per day}} = \text{no. of days}$

$$\frac{50 \text{ strips}}{4 \text{ strips per day}} = \mathbf{12.5 \text{ days}}$$

5. $°F = \dfrac{9}{5}°C + 32$

$$°F = \frac{9}{5}(39) + 32$$

$$\mathbf{°F = 102.2 \text{ degrees}}$$

6. $\dfrac{1 \text{ pound}}{0.45 \text{ kilograms}} = \dfrac{48 \text{ pounds}}{k \text{ kilograms}}$

$$k \times 1 = 0.45 \times 48$$

$$\mathbf{k = 21.6 \text{ kilograms}}$$

7. $\dfrac{20 \text{ mg}}{1 \text{ kg}} = \dfrac{n \text{ mg}}{21.6 \text{ kg}}$

$$n \times 1 = 20 \times 21.6$$

$$\mathbf{n = 432 \text{ mg}}$$

8. $\dfrac{5 \text{ ml}}{10 \text{ mg}} = \dfrac{m \text{ ml}}{5 \text{ mg}}$

$$10 \times m = 5 \times 5$$

$$10m = 25$$

$$\mathbf{m = 2.5 \text{ ml}}$$

9. $\dfrac{2 \text{ mg}}{1 \text{ ml}} = \dfrac{0.6 \text{ mg}}{n \text{ ml}}$

$$2n = 1 \times 0.6$$

$$\mathbf{n = 0.3 \text{ ml}}$$

10. *convert:* $0.5 \text{ g} = 500 \text{ mg}$

$$\frac{500 \text{ mg}}{1 \text{ ml}} = \frac{750 \text{ mg}}{n \text{ ml}}$$

$$500 \times n = 1 \times 750$$

$$n = \frac{750}{500}$$

$$\mathbf{n = 1.5 \text{ ml}}$$

Pages 172–173

1. $\dfrac{6 \text{ miles}}{1 \text{ gallon}} = \dfrac{1,410 \text{ miles}}{n \text{ gallons}}$

$$6 \times n = 1 \times 1,410$$

$$6n = 1,410$$

$$\mathbf{n = 235 \text{ gallons}}$$

2. flat fee + rate × miles = delivery charge

$$200 + \$5 \times 52 = \text{delivery charge}$$

$$\mathbf{\$460} = \text{delivery charge}$$

3. $\dfrac{1 \text{ liter}}{0.26 \text{ gallon}} = \dfrac{600 \text{ liter}}{n \text{ gallons}}$

$$1 \times n = 0.26 \times 600$$

$$\mathbf{n = 156 \text{ gallons}}$$

4. $\dfrac{1 \text{ mph}}{1.61 \text{ kph}} = \dfrac{65 \text{ mph}}{n \text{ kph}}$

$$1 \times n = 1.61 \times 65$$

$$\mathbf{n = 104.65 \text{ kph}}$$

5. Sal's to Catalpa + Catalpa to Southview + Southview to Sal's = shortest route

$$8.9 \text{ mi} + 15.1 \text{ mi} + 11.9 \text{ mi} = \text{shortest route}$$

$$\mathbf{35.9 \text{ miles}} = \text{shortest route}$$

6. Sal's to Oak Grove + Oak Grove to High Mountain + High Mountain to Oak Grove + Oak Grove to Outlook Falls + Outlook Falls to Southview + Southview to Catalpa + Catalpa to Sal's = shortest route

$$14.7 \text{ mi} + 8.2 \text{ mi} + 8.2 \text{ mi} + 9.8 \text{ mi} + 12.2 \text{ mi} + 15.1 \text{ mi} + 8.9 \text{ mi} = \text{shortest route}$$

$$\mathbf{77.1 \text{ mi}} = \text{shortest route}$$

7. Sal's to Southview + Southview to Outlook Falls + Outlook Falls to Southview + Southview to Sal's = shortest route

$$11.9 \text{ mi} + 12.2 \text{ mi} + 12.2 \text{ mi} + 11.9 \text{ mi} = \text{shortest route}$$

$$\mathbf{48.2 \text{ mi}} = \text{shortest route}$$

Pages 175–176

1. $\dfrac{200 \text{ cubic feet of water}}{1 \text{ hour}} = \dfrac{\text{total cubic feet of water}}{n \text{ hours}}$

$$8 \text{ inches} = \frac{2}{3} \text{ foot}$$

$$30 \text{ ft} \times 40 \text{ ft} \times \frac{2}{3} \text{ ft} = \text{total cu ft of water}$$

$$800 \text{ cu ft} = \text{total cu ft of water}$$

$$\frac{200 \text{ cubic feet of water}}{1 \text{ hour}} = \frac{800 \text{ cubic feet of water}}{n \text{ hours}}$$

$$200n = 800$$

$$\mathbf{n = 4 \text{ hours}}$$

2. water pumped out − water in basement =
water seeped in
water pumped out = 300 cubic feet per hour ×
6 hours = 1800 cubic feet
9 inches = $\frac{3}{4}$ foot
water in basement = 30 feet × 50 feet × $\frac{3}{4}$ foot
= 1125 cubic feet
1800 cubic feet − 1125 cubic feet = **675 cubic feet**

3. The shortest route is **4.8 miles.** One way to get
this total is to go to the gas station, then the
video store, the department store, the library,
the grocery, and finally home.

4. $\frac{\text{number of min}}{\text{change in sec}} = \frac{\text{number of min}}{\text{change in sec}}$

$\frac{10 \text{ min}}{5 \text{ sec change}} = \frac{n \text{ min}}{10 \text{ sec change}}$

$5n = 100$

$n = \textbf{20 minutes}$

5. See if you can discover why a square would have
the smallest possible perimeter.
$A = lw = l \times l$
$400 = l \times l$
$l = \textbf{20 feet.}$ The dimensions of the garden
should be 20 feet by 20 feet.

6. $P = 2l + 2w$
$100 = 2l + 2(3)$
$100 = 2l + 6$
$94 = 2l$
47 feet $= l$
The run could be 47 feet long.

7. convert miles per hour to feet per second

$\frac{95 \text{ miles}}{1 \text{ hour}} \times \frac{5,280 \text{ ft}}{1 \text{ mile}} = \frac{501,600 \text{ ft}}{1 \text{hour}}$

$\frac{501,600 \text{ ft}}{1 \text{ hour}} \times \frac{1 \text{ hour}}{60 \text{ min}} = \frac{8,360 \text{ ft}}{1 \text{ min}}$

$\frac{8,360 \text{ ft}}{1 \text{ min}} \times \frac{1 \text{ min}}{60 \text{ sec}} = \frac{139 \text{ ft}}{1 \text{ sec}}$

proportion: $\frac{139 \text{ ft}}{1 \text{ sec}} = \frac{60.5 \text{ ft}}{n \text{ sec}}$

$139n = 60.5$

$n = \frac{60.5}{139} = \textbf{0.44 sec}$

(rounded to the nearest hundredth)

Pages 177–182, Posttest B

1. c. *conversion:* 1 dozen eggs = 12 eggs

$\frac{1\frac{1}{2} \text{ pounds}}{12 \text{ eggs}} = \frac{n \text{ pounds}}{8 \text{ eggs}}$

$12 \times n = 1\frac{1}{2} \times 8$

$n = \frac{3}{2} \times 8 \times \frac{1}{12} = \frac{24}{24} = \textbf{1 pound}$

2. c. original cost per socket set − price of last set
= money lost on last set
missing information:
total cost for sets ÷ number of sets =
original cost per tool set
540.60 ÷ 15 tool sets = $36.04
$36.04 − $24.00 = **$12.04**

3. c. weekend's hot dogs − Saturday's = Sunday's
hot dogs
426 hot dogs − 198 hot dogs = **228 hot dogs**

4. b. $\frac{\text{part}}{\text{whole}} = \frac{\text{percent}}{100}$

$\frac{156,000}{n} = \frac{39}{100}$

$39 \times n = 156,000 \times 100$

$n = 15,600,000 \div 39 = \textbf{400,000 votes}$

5. b. passengers per jet × number of jets = total
passengers
214 passengers × 96 jets = **20,544 passengers**

6. b. tile − size needed = size cut off
$\frac{3}{4}$ foot − $\frac{1}{3}$ foot = $\frac{5}{12}$ **foot**

7. e. not enough information given. You need to
know how many apples were in the bushel.

8. c. $\frac{\text{concrete}}{\text{square feet}} = \frac{\text{concrete}}{\text{square feet}}$

$\frac{1.23 \text{ cubic yards}}{100 \text{ square feet}} = \frac{n \text{ cubic yards}}{550 \text{ square feet}}$

$100 \times n = 1.23 \times 550$

$n = 676.50 \div 100 = 6.765 \text{ cubic yards}$

$n = \textbf{6.77 cubic yards}$

9. c. pounds for recipe − pounds in freezer =
pounds needed
4 pounds − 2.64 pounds = **1.36 pounds**

10. c. $\frac{\text{gallons}}{\text{people}} = \frac{\text{gallons}}{\text{people}}$

$\frac{1,638,000 \text{ gallons}}{78,000 \text{ people}} = \frac{n \text{ gallons}}{1 \text{ person}}$

$n \times 78,000 = 1 \times 1,638,000$

$n = 1,638,000 \div 78,000$

$n = \textbf{21 gallons}$

11. e. total weekly earnings × number of weeks =
gross yearly earnings
missing information:
money taken out + take-home pay = total
weekly earnings
198.23 + 532.77 = $731.00
731.00 × 52 weeks = **$38,012.00**

12 a. closed stations + remaining stations = last year's stations
423 stations + 2,135 stations = **2,558 service stations**

13. a. fraction (of) × total representatives = votes
$\frac{2}{3} \times 435 =$ **290 votes**

14. e. not enough information given. You need to know how much she deposited.

15. b. inches for door + inches for quarter panel = inches needed
$33\frac{3}{4}$ inches + $51\frac{2}{3}$ inches = $85\frac{5}{12}$ **inches**

16. d. weight of roast × price per pound = total price
2.64 pounds × 3.96 = $10.4544 = **$10.45**

17. c. $\frac{432 \text{ male workers}}{\text{total work force}} = \frac{\text{percent male workers}}{100}$

missing information:
100% work force − percent female = percent male
100% − 28% = 72% male

$\frac{432 \text{ male workers}}{n \text{ workers}} = \frac{72}{100}$
$72 \times n = 432 \times 100$
$n = 43,200 \div 72 =$ **600 workers**

18. c. weight of carton ÷ number of nails = weight per nail
$\frac{3}{4}$ pound ÷ 75 nails = $\frac{3}{4} \times \frac{1}{75} = \frac{1}{100}$ pound = **0.01 pound**

19. a. $\frac{\text{part}}{\text{whole}} = \frac{\text{percent}}{100}$
$\frac{7 \text{ field goals}}{25 \text{ field goal attempts}} = \frac{n}{100}$
$25 \times n = 7 \times 100$
$n = 700 \div 25 =$ **28%**

20. d. original weight + first month + second month = new weight
104 pounds + 3 pounds + 4 pounds = **111 pounds**

21. a. roast + steak = total meat
3.69 pounds + 1.23 pounds = **4.92 pounds**

22. b. original balance − total checks + deposit = new balance
missing information:
first check + second check = total checks
46.19 + 22.45 = $68.64
$74.81 − $68.64 + $60.00 = **$66.17**

23. d. $\frac{\text{part}}{\text{whole}} = \frac{\text{percent}}{100}$
$\frac{n \text{ pounds}}{450 \text{ pounds}} = \frac{9}{100}$
$100 \times n = 9 \times 450$
$n = 4,050 \div 100 =$ **40.5 pounds**

24. b. $\frac{\text{miles}}{\text{gallon}} = \frac{\text{miles}}{\text{gallon}}$
$\frac{168 \text{ miles}}{5.6 \text{ gallons}} = \frac{417 \text{ miles}}{n \text{ gallons}}$
$168 \times n = 5.6 \times 417$
$n = 2,335.2 \div 168 =$ **13.9 gallons**

25. e. $\frac{\$}{\text{pound}} = \frac{\$}{\text{pound}}$
$\frac{\$4.79}{1 \text{ pound}} = \frac{\$2.06}{n \text{ pounds}}$
$\$4.79 \times n = \$2.06$
$n = \$2.06 \div \$4.79 =$ **0.43 pound**

FORMULAS

PERIMETER

Figure	Name	Formula	Meaning
	Rectangle	$P = 2l + 2w$	l = length w = width
	Square	$P = 4s$	s = side

AREA

Figure	Name	Formula	Meaning
	Rectangle	$A = lw$	l = length w = width
	Square	$A = s^2$	s = side

VOLUME

Figure	Name	Formula	Meaning
	Rectangular solid	$V = lwh$	l = length w = width h = height
	Cube	$V = s^3$	s = side

CONVERSIONS

Time

365 days = 1 year
12 months = 1 year
52 weeks = 1 year
7 days = 1 week
24 hours = 1 day
60 minutes = 1 hour
60 seconds = 1 minute

Length and Area

5,280 feet = 1 mile
1,760 yards = 1 mile
3 feet = 1 yard
36 inches = 1 yard
12 inches = 1 foot
144 square inches = 1 square foot
4,840 square yards = 1 acre
1,000 meters = 1 kilometer
100 centimeters = 1 meter
1,000 millimeters = 1 meter
10 millimeters = 1 centimeter

Weight	Mass
2,000 pounds = 1 ton	1,000 grams = 1 kilogram
16 ounces = 1 pound	1,000 milligrams = 1 gram

Volume

4 quarts = 1 gallon
2 pints = 1 quart
4 cups = 1 quart
2 cups = 1 pint
32 ounces = 1 quart
16 ounces = 1 pint
8 ounces = 1 cup
1,000 milliliters = 1 liter

Metric to Customary

1 kilometer = 0.62 mile
1 meter = 39.37 inches
1 centimeter = 0.39 inch
1.61 kilometers per hour = 1 mile per hour
1 liter = 1.057 quarts
1 liter = 0.264 gallon
1 kilogram = 2.2 pounds
1 gram = 0.035 ounce

GLOSSARY

A

approximation An estimate that is close to a given number, but not exact. A bag of fruit marked $2.89 will cost approximately $3.00.

arithmetic operations Addition, subtraction, multiplication, and division.

C

combination word problem A word problem that needs 2 or more steps to be solved.

conversion Changing from one type of measurement to another. 12 inches = 1 foot

D

denominator The bottom number of a fraction. In the fraction $\frac{1}{2}$, 2 is the denominator.

diagram A picture or visual aid that helps you understand a word problem.

E

estimate An approximate amount used to determine the accuracy of the arithmetic, or used to give an idea of the answer before doing the arithmetic. 17 + 52 can be rounded to 20 + 50, so the estimate is 70.

F

formula An equation that states a rule or factual information that can be used to solve a certain type of problem. $A = lw$ is the formula for finding the area of a rectangle.

G

given information All the numbers and labels that are in a word problem.

K

key word A clue that can help you decide which arithmetic operation to use. In the problem find the difference between 15 and 6, the key word *difference* helps you decide to subtract.

L

label The noun (word or symbol) that a number refers to. If you are adding 6 apples and 5 apples, the total amount will be 11 and the label will be apples.

M

math intuition A general understanding of numbers, math operations, and a feel for what a solution should be.

mental math Arithmetic operations that can be done in your head, without the use of pencil and paper or calculator.

N

necessary information The numbers and labels in a word problem that are needed to find a solution.

number sentence A restatement of a word problem as an equation using numbers and labels. Sample problem: Jody spent $4.29 and Doretha spent $6.45 for lunch. Find the total amount spent. The number sentence would be: $4.29 + $6.45 = $ total amount

numerator The top number of a fraction. In the fraction $\frac{3}{4}$, 3 is the numerator.

O

order of operations Rules that govern the sequence when you must use more than one arithmetic operation. Order of operations: (1) do multiplication and division before addition and subtraction; (2) do arithmetic in parentheses first; (3) compute from left to right.

P

part A piece that is being compared to a whole.

percent A part of a hundred.

percent circle A memory aid used to help solve percent problems.

proportion A math equation that states that two ratios are equal. $\frac{2}{4} = \frac{1}{2}$ is a proportion.

Q

question The part of a word problem that tells you what to look for.

R

ratio A comparison of the relative size of two groups. The ratio for 3 teachers working with 15 students is 3 teachers:15 students.

restating a problem Saying a problem in your own words to help understand what the problem is asking.

round Estimate to a particular place value. The number 779 rounded to the nearest hundred is 800; rounded to the nearest ten is 780.

S

solution A number and label that will correctly answer the question.

solve To find the solution.

substitution Temporarily replacing a difficult-to-understand number with a small whole number that is easier to picture.

W

word problem A sentence or group of sentences that tells a story, contains numbers, and asks the reader to find another number.

whole A complete amount; in percent, the base for comparison.

INDEX